YOUR TICKET
TO THE
UNIVERSE

YOUR TICKET TO THE UNIVERSE

A GUIDE TO EXPLORING THE COSMOS

KIMBERLY K. ARCAND and MEGAN WATZKE

Foreword by
MARIO LIVIO

SMITHSONIAN BOOKS WASHINGTON, DC

CONTENTS

A crescent moon sets behind the thin blue line of Earth's atmosphere. An International Space Station (ISS) crewmember photographed this image as the ISS spacecraft passed over central Asia.

On August 27, 2011, astronaut Ron Garan aboard the International Space Station (ISS) photographed one of the sixteen sunrises that astronauts get to see each day. This photo shows the Sun rising as the ISS flew between Rio de Janeiro, Brazil, and Buenos Aires, Argentina.

IN THE MOVIE *NOTTING HILL*, BRITISH ACTOR HUGH GRANT PLAYS WILLIAM, the timid owner of a little travel bookshop. One day, Anna Scott (played by Julia Roberts), supposedly the biggest movie star in the world, walks into his modest shop, looking for a book on Turkey. William recommends a particular travel guide to her because, he says, "The man who wrote it has actually been to Turkey, which helps." While Kimberly Arcand and Megan Watzke clearly have not been to the farthest reaches of the cosmos, they have both been intimately involved with explorations of the universe by NASA's Chandra X-ray Observatory. As such, they make for superb guides for the organized cosmic tour represented by this book. The metaphorical "rooms" they book for you as you travel from Earth to the most remote parts of the universe always have a spectacular view. The book, lavishly illustrated with actual images taken by the Hubble, Chandra, and Spitzer space telescopes, as well as images from other NASA spacecraft and a variety of ground-based telescopes, uses the entire electromagnetic spectrum to give a complete panorama. The explanations provided by your guides are always fresh, clear, and spiced with humor and references to current popular culture.

The tour starts on Earth, then visits the Moon and the Sun before stopping at all the major attractions in our Solar System. Your guides then put you on a faster space ship and take you to the stars, viewed at various stages of their evolutionary lives. It is then that the sweeping cosmic tour begins, from our own Milky Way galaxy to other galaxies near and far, placid and violently colliding. Along the way, your guides allow you to take snapshots of the dynamic vicinity of black holes, experience the gravitational effects of dark matter, and ponder the mysterious cosmic acceleration caused by dark energy.

As the authors admit, we scientists do not understand all of the phenomena that you will encounter on this trip. On the contrary, they emphasize how much is left to be explored in this vast cosmic landscape. This is an exhilarating tour—so fasten your seat belt, sit back, relax, and enjoy the amazing scenery. Dinner is not provided, but there's plenty of food for thought!

— MARIO LIVIO

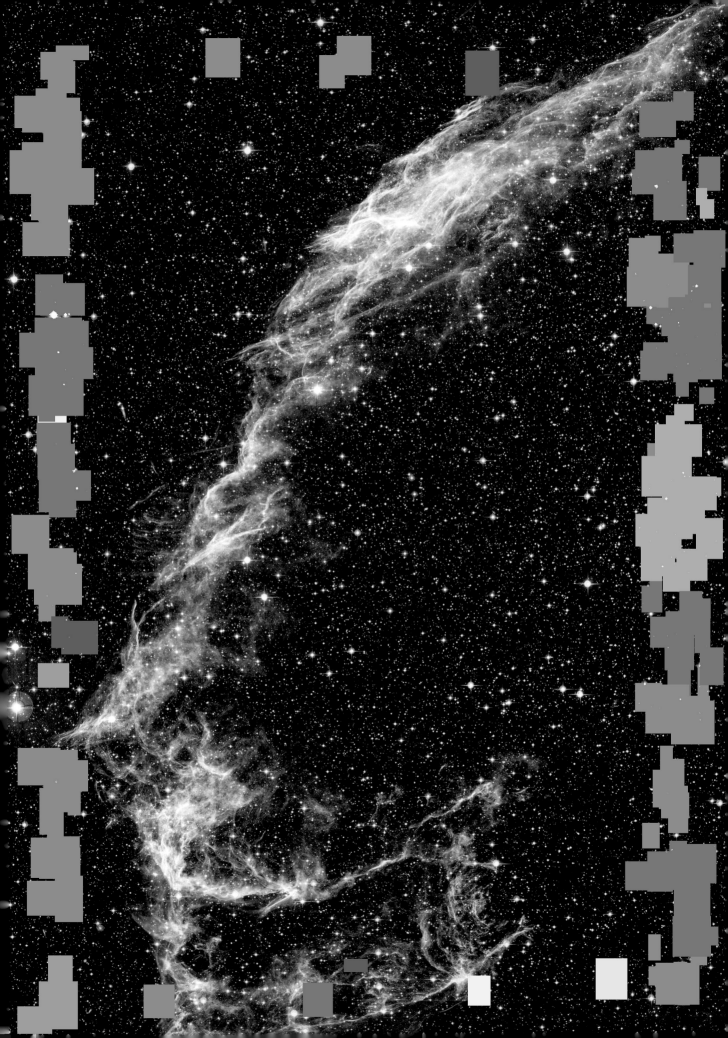

The sky belongs to everyone. That's the premise of this guidebook to the Universe. You don't need a medical degree to know when you're sick or a doctorate in literature to appreciate a novel. In the same spirit, even those of us who do not have advanced degrees in astronomy, astrophysics, or space science can gain access to all the wonder and experience that the Universe has to offer.

The goal of this book is to get you cosmically oriented for your own exploration, guiding you through the Universe, step by step, with pictures along the way to show where we're going and to point out must-see sights that no celestial traveler should miss. We might leave out someone's favorite galaxy or a famous nebula, but that's the nature of a travel guide. We'll start our journey locally on Earth, hit our favorite star (the Sun, that is), head out through our Solar System, and then travel far, far beyond it.

The more we look at the Universe, the more interesting it gets. In recent years, astronomers have learned more about black holes, found hundreds of planets around other stars, and determined that 96 percent of the Universe is made up of stuff that we haven't yet been able to figure out. Everything we know about the Universe comes from basic and applied science, even if some of it may sound as if it comes from science fiction.

Welcome to your Universe.

– MEGAN and KIM

OPPOSITE:

The Veil Nebula represents the remains of a star that exploded possibly between 5,000 and 8,000 years ago in our Milky Way galaxy. The original supernova was probably as bright as a crescent Moon, and people who had just invented the wheel and writing could have viewed it for weeks. The Veil Nebula is a large relatively faint supernova remnant that has since expanded to cover an area of about six times the diameter of a full moon. Here we can see the gas between the stars being heated by the incredible blast wave that is still expanding through space.

1

YOU ARE HERE

Begin at the beginning and go on till you come to the end; then stop.

— LEWIS CARROLL

BEFORE WE CAN EXPLORE BEYOND OUR PLANET, we need to gather some information from our own experiences living here on Earth. This is not just for the sake of nostalgia. As we learn more about our Solar System, the Milky Way, and ultimately the Universe, we find there are valuable tools for a cosmic explorer's toolkit right here on our own home planet.

Most humans have never left the planet—only a few hundred people have ever gotten into space, including astronauts and space tourists, compared to the billions of people who live on the planet full-time. Our Earthbound nature introduces the possibility, though, of misconceptions about what is out there, beyond our planet. For example, what if gravity acts differently in other environments? For decades, scientists have been working very hard on this question—including Albert Einstein, who famously made major progress on it.

PREVIOUS SPREAD:

Our own home planet offers a rich opportunity to expand our understanding of life.

A REAL GENIUS

There's a reason why Albert Einstein's last name is often used as a synonym for "genius." There probably is no other individual in history who has made as much of an impact on our understanding of the Universe as Einstein (1879–1955), the German-born physicist who is most famous for his theory of relativity. When you realize that he was coming up with ideas and theories decades before we had the telescopes or other facilities needed to test them, it's truly amazing.

Einstein was able to think about the Universe on scales that were way ahead of his time. He realized that scientists would need to account for the enormous masses and scales of the cosmos as well as the extreme environments that might exist in the Universe—and he did just that. To this day, more than a hundred years after his first batch of watershed papers was published, scientists still rely on Einstein's ideas to process what we see across the Universe.

Nowhere is our potential bias more important than in the search for life outside of our planet. For a long time, most scientists were pretty convinced that any type of life would need what we need: liquid water, something decent to breathe, and a comfortable environment. Today, we know that Earth is just one of probably billions of planets in our own Galaxy—and the Milky Way is just one among a hundred billion other galaxies, or even more. At the same time, we have learned that the recipe for life on Earth may be more diverse than anyone used to think. This means that the possibilities for life elsewhere in the Universe are likely to be much greater than previous generations of scientists thought.

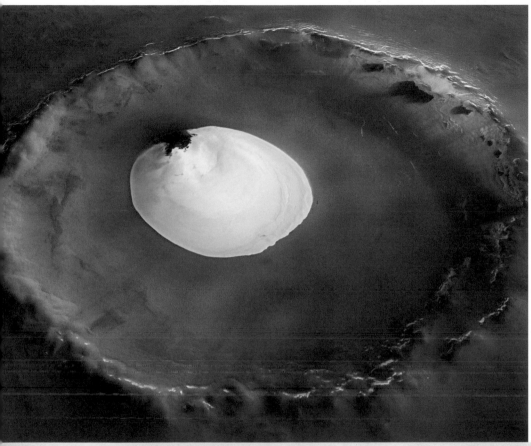

Earth or not? The left image shows an impact crater on Mars. The bright patch of material near the center of the 22-mile-wide (35 kilometer) crater is residual water ice. The image below is an aerial photograph of Pingualuit Crater in northern Quebec, Canada. The crater is a little over two miles (3.2 kilometers) in diameter. Scientists estimate that it formed about 1.4 million years ago.

A COSMIC PRIMER:
LIFE ON EARTH

What is "life," after all? Some of us might define life using examples, such as humans and animals, plus plants and smaller things like bacteria and microbes. Most of these things exist comfortably right alongside us in our everyday lives, or they live in environments that we can visit, such as oceans, forests, tundra, and so on.

In recent years, however, scientists have begun seeking out some of the harshest, seemingly inhospitable spots on the planet to look for life. They have explored volcanic vents thousands of feet under the ocean, where the pressure is so intense it can crush even the strongest submarine. They have looked in frozen lakes encapsulated under miles of Antarctic ice, where sunlight has not shone for thousands of years. They have checked inside highly acidic, salty mountain lakes thousands of feet high, where the ultraviolet radiation from the Sun bombards anything living there.

Wherever they have gone, scientists have found life at just about every turn.

Researchers are looking at these unwelcoming sites not only to see what can exist on Earth, but what can exist beyond it. Decades ago, many scientists would have insisted that life needed certain key ingredients and particular environments. Now, however, our thinking has expanded to match the latest discoveries. If life can thrive under extreme pressures several miles below the surface of an ocean on Earth, then why could it not thrive in some other equally harsh environment elsewhere in space? If life exists under an ice-encrusted lake in Antarctica, might it be found under the surface of an ice-encrusted moon of Jupiter?

The ultimate scenario for researchers would be to send probes—and perhaps, one day, scientists themselves—to the far-flung corners of the Solar System to explore and look for life. Today, even sending robotic probes remains very difficult and expensive to do, so such missions are few and infrequent. To get around that problem, scientists and engineers have turned to remote locations on Earth to learn as much as possible about exotic environments.

It is hard to imagine Earth being used as a mere substitute for another planet, in the way that a Hollywood backdrop is a substitute for an actual building, but having these "planetary doubles" on Earth is extremely important for people who want to learn about places away from our home planet. These sites are used to test not only what types of life may be able to live in less hospitable environments, but also what types of challenges that vehicles and other mission hardware would experience if they were sent to such foreign places. Earth plays an important role, then, in helping us understand what lies beyond it.

The ice covering Lake Fryxell in Antarctica is a result of melting glaciers. Freshwater stays on top of the lake and freezes, sealing in salty water deep below.

Let's take this new (and certainly more complicated) understanding of life here on Earth, pack up our things, and head out to the rest of the cosmos. Just bear in mind as we make our way: The more we learn, the less we sometimes understand.

A HOLISTIC VIEW OF SCIENCE

From an early age, we are taught that science is split into different categories: biology, chemistry, physics, and so on. As we get farther along in school, this fragmentation seems to accelerate. Biology isn't just "biology" anymore—it's molecular biology, cellular biology, systems biology, evolutionary biology, and so forth. Chemistry splinters into inorganic chemistry, atmospheric chemistry, materials chemistry, and so on. Once you get to college or graduate school, sometimes these sub-disciplines seem as different as night and day.

All of science, however, is connected and governed by the same basic laws. The physics between two billiard balls is the same as those between two planets. There are certainly some complexities and differences, but the underlying principles are the same.

The study of the Universe is particularly good at bringing together different fields of science here on Earth. Astronomy and physics are completely entwined with chemistry, ranging from how stars burn to how galactic winds disperse elements into space. Astronomy also has a natural connection with geology (studying other planets and bodies in our Solar System and beyond), biology (trying to figure out where and how life can exist), and many others. The bottom line is this: All science serves us in the quest to discover and understand the unknown.

There is no true dividing line between our planet's atmosphere and the rest of space. Scientists think of the altitude about 60 miles (100 km) above the Earth's surface as the "top of the atmosphere," pictured here.

YELLOWSTONE NATIONAL PARK

Life creates a beautiful array of colors in a hot spring in Yellowstone National Park, at the place where the states of Wyoming, Montana, and Idaho come together. Scientists have found that many different types of life (known as microorganisms) live in the pools there. Because the water in the springs can be extremely hot, these tiny organisms, some made of single cells and others of multiple cells, are called "extremophiles." Different extremophiles thrive in different temperatures, and the color of a particular area is determined by which organisms are living in it. By studying extremophiles here on Earth, scientists are trying to get a handle on where life might thrive in similar environments elsewhere in the Solar System.

SHARK BAY

For about 85 percent of the history of life on Earth, only tiny organisms known as microbes existed. The only large-scale evidence of their activities is preserved by stromatolites, which are rocky, dome-shaped structures that form in shallow water. They build up over the years by adding layers that include the microbes that live there. These bio-buildups mostly occur in lakes and lagoons where extreme conditions—such as very high salt levels—prevent animals from grazing there. One such location where stromatolites have survived is the Hamelin Pool Marine Nature Reserve in Shark Bay, Western Australia. Stromatolites give scientists a chance to study what the world was like when only microbes existed on Earth.

BLUE HOLES OF THE BAHAMAS

For more than a billion years when our planet was very young, Earth's oceans were without oxygen. The microorganisms that lived in them consumed light through the process of photosynthesis, but they did not produce oxygen. Today, in flooded caves called "blue holes" found on islands in the Bahamas, scientists study dense blooms of purple and green bacteria that also harvest light without producing oxygen. The lifestyle contrasts with that of modern plants and certain bacteria that produce oxygen as a by-product of photosynthesis, providing the oxygen-rich atmosphere that supports humans and other life on the planet. A team of scientists and expert cave divers captured this image of a blue hole during a National Geographic expedition.

SVALBARD

Off the coast of northern Norway, deep within the Arctic Circle, lies a remote archipelago known as Svalbard. Because of eruptions that happened under the glacier about a million years ago, this region has a unique combination of volcanoes, hot springs, and permafrost (when soil is frozen for at least two years or longer). Svalbard provides a great opportunity to study the interaction between water, rocks, and primitive life forms in a Mars-like environment. Scientists travel here to test the protocols, procedures, and equipment needed to detect traces of the elements necessary for life on Mars.

HIGH LAKES, LAGUNA VERDE

Volcanic lakes in the highest mountain ranges on Earth, such as the Andes of South America, make excellent analogies for lakes that scientists think existed on Mars about 3.5 billion years ago. This includes Bolivia's Laguna Verde, at an elevation of some 14,240 feet, (4,340 meters), pictured here. Astrobiologists (that is, scientists who study the origin, evolution, and distribution of life around the Universe) are studying the impact of rapid climate change on this lake and ones like it, as well as their ability to sustain life in a changing environment. The results could shed light on the fate of life that may have arisen on Mars in the past and become extinct as the planet's geology and climate evolved.

SIMBA LAKE

Like Laguna Verde, SImba Lake is located high in the Andes Mountains. Simba Lake is red because of algae that developed pigments to protect themselves against the high levels of ultraviolet radiation found at nearly 20,000 feet (6,100 meters) above sea level. Researchers working on Simba Lake study the impact of rapid climate change on lake habitat and life's adaptability in an effort to understand the evolution of the early environments of both Mars and Earth.

FOUR MARSHES

This image shows one of the many gorgeous pools that make up the biological reserve of Cuatro Ciénegas ("four marshes") located in the desert of the Mexican state of Coahuila. In addition to the extreme diversity of plant and animal life protected by the reserve, many of the pools contain living colonies of tiny organisms, which provide records of ancient life on Earth. They provide researchers with a living laboratory in which to study the possible origins of life on Earth.

RIO TINTO

The Rio Tinto in southwestern Spain wanders for 60 miles (97 kilometers) before joining the Atlantic Ocean. Despite its acidic waters and high concentrations of iron and other heavy metals, the river supports an incredible diversity of extremophile microorganisms, including algae and fungi. Microbes and the surfaces they stick to colonize the riverbed and are covered with yellow iron oxide. Because of geological similarities with Mars, scientists tested equipment at Rio Tinto in 2005 for drilling on Mars in search of life below the planet's surface.

SANDSTONE RECORDS

This seemingly innocuous piece of upturned sandstone in South Australia actually shows ripple marks of an ancient seabed. This part of Australia hosts fossils of the first complex multicellular organisms on Earth, which began to emerge about 600 million years ago. The study of these early fossils helps scientists learn more about how complex life arose and evolved on Earth and how it might evolve on other planets.

MONO LAKE

Mountains surround California's Mono Lake, an inland sea of about 700 square miles (1,800 square kilometers) located just east of Yosemite National Park. Mono Lake is a closed hydrological basin, which means that water flows in but does not flow out. Since the only way for water to leave Mono Lake is through evaporation, it is twice as salty as the water of the ocean. Freshwater streams and underwater springs have brought trace amounts of minerals, including arsenic, into the lake over the millennia. Scientists are studying bacteria here to understand how life has evolved in this environment.

2

OUR EARTH
IN SPACE

The whole of science is nothing more than a refinement of everyday thinking.
— ALBERT EINSTEIN

U NDERSTANDING LIGHT IS THE INITIAL STEP IN ANY TRIP across the cosmos. That's because light is the equivalent to transportation to sites throughout the Universe. We can't fly to these far-off places just yet, but we can learn about and explore these exotic destinations by the light they give off.

We know about light by living here on Earth. It seems like most people know the difference between light and dark, right? Well, just to be sure, ask yourself to define what light actually is. What many people think of as light—that is, the stuff we detect with our eyes—is just a very narrow sliver of the full gamut of light that exists.

You might already be familiar with other types of light, perhaps without being fully aware of them. The night vision goggles used by the military, for example, can "see" in the dark because they detect heat from people and other things in the form of infrared radiation. (Radiation is really just another term for light, and we will use the two words interchangeably.) You are also likely familiar with ultraviolet or UV light, which is radiation that can cause damage to our skin and why we buy sunscreen to block it. If you have been to the dentist—or, unluckily, broken a bone—you have experienced an even more energetic type of light in the form of an X-ray.

All light travels at the same speed. But there are different kinds of light, each with its own range of energies. Radio waves are the lowest energy of the spectrum. Visible light, also called optical light, is the only light we can see with the human eye. It is a million times more energetic than the average radio wave.

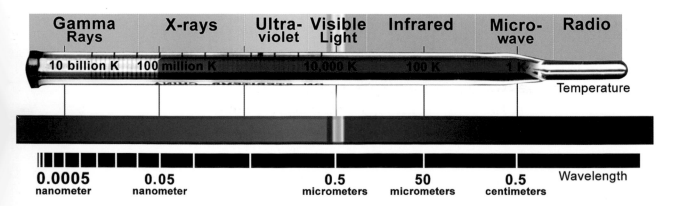

| Gamma Rays | X-rays | Ultra-violet | Visible Light | Infrared | Micro-wave | Radio |

| 10 billion K | 100 million K | | 10,000 K | 100 K | 1 K | |

Temperature

| 0.0005 nanometer | 0.05 nanometer | | 0.5 micrometers | 50 micrometers | 0.5 centimeters | Wavelength |

The wavelength of radiation produced by an object is usually related to its temperature.

The energy of X-ray light can range from hundreds to thousands of times higher than that of visible light.

We will need to take this knowledge of "other" types of light with us when we go exploring the cosmos, because not only do objects in the Universe glow in radiation that we cannot see with our eyes, but they also actually give off most of their light in the range that is invisible to us without the help from telescopes or other instruments.

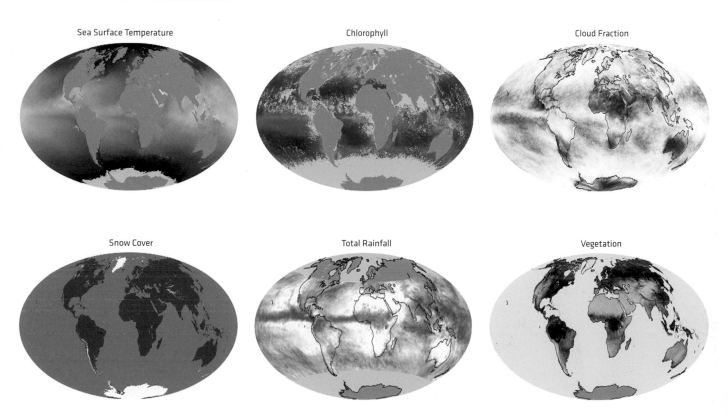

Sea Surface Temperature   Chlorophyll   Cloud Fraction

Snow Cover   Total Rainfall   Vegetation

Scientists use different kinds of light to study different things. Take the Earth, for example. We use satellites that can detect microwaves to study sea temperatures, as well as soil moisture, sea ice, ocean currents, and pollutants. Scientists use infrared radiation to examine the thickness of ice in the polar regions, to help study volcanic eruptions, and to measure vegetation cover on our planet. Researchers use optical light to measure Earth's snow cover. You'll notice in the snow cover image above that it appears no snow occurred at the highest latitudes of the Northern hemisphere. That is because we cannot collect useful optical data during winter when no sunlight reaches those regions. So, in short, it takes different sets of eyes in the sky to get a comprehensive view of what's happening on our planet.

NASA's Earth-viewing satellites provide a global look at what's happening right here on our planet.

DO WE REALLY NEED ALL THESE TELESCOPES?

Throughout this book, you will find references to many different telescopes on the ground and in space. Why do astronomers need to build all of these?

It might seem like a wealthy person who collects a fleet of fancy cars in his or her garage but really needs only one to drive. But the reality is that each professional telescope has unique capabilities that set it apart from the others. Modern professional telescopes are complex pieces of technology. It takes a lot of scientific, engineering, and technical know-how to detect a certain type of infrared light or a sliver of gamma rays.

A good analogy is the need for cars, planes, trains, and boats in traveling. Depending on where you are going, you might need one or more modes of transportation. Each serves a different purpose—a car cannot be expected to get you over an ocean, and a boat will not be very helpful in crossing continents. The same is true of different kinds of telescopes: Each has its own capabilities, and one cannot necessarily replace another.

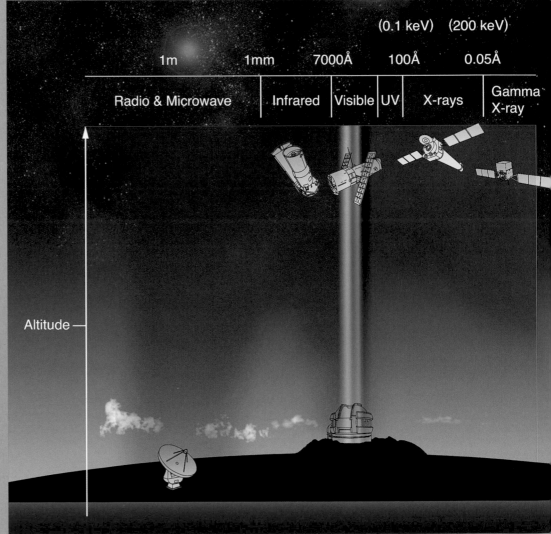

Different kinds of telescopes are needed to look at different kinds of light. This illustration shows which kinds of light are absorbed by Earth's atmosphere. For example, X-rays (shown toward the upper right) are blocked by the atmosphere, so we must launch spacecraft into orbit to study them.

TAKING AN ALIEN TO A BASEBALL GAME

As we have seen, the light we can detect with our eyes reveals only a small fraction of what is going on in the Universe. To illustrate this point, we'll use a nonscientific analogy from one of our favorite resources, the orbiting telescope called the Chandra X-ray Observatory (chandra.si.edu).

Imagine that an alien has landed on our planet. It's your job to bring him/her/it directly to a baseball game—blindfolded—and to get the alien to understand this game. You lift the alien's blindfold, but for some reason, the alien can only see a sliver down the third base line. From this viewpoint, you ask your new alien friend to tell you not only the score, but also the number of players, the rules, and so on. Our poor extraterrestrial visitor would have a tough time working it out—and would probably look for some excuse to leave. If the alien could see the whole field, however, then he/she/it would have a better chance of figuring out the game.

It would be hard to figure out the rules of baseball if you could only see the third base line.

The analogy to studying space is that visible light represents only looking down the third base line. All of the other types of light, from radio waves to gamma rays, fill in the rest of the field. As is certainly true of baseball, it's still tricky to understand all of the rules of the Universe, but it's a lot easier when we can get a full picture.

Light is much more than we can see with our eyes. It includes everything from radio waves to gamma rays—most of which is invisible without technology to "see" it. Objects in the Universe—from our Sun to distant galaxies—give off most of their light in the invisible regime. Colors are assigned to astronomy data to make images so that we can see and interpret the information.

X-ray image of the center of our Galaxy from the Chandra X-ray Observatory.

COLORFUL WORLDS

The next concept that we will look at is color. Here is another case where being residents of planet Earth gives us a head start. Many preschoolers learn about the colors of the rainbow: red, orange, yellow, green, blue, indigo, and violet (or "Roy G. Biv" as it's sometimes called). As we get a little older, a science teacher may show us how sunlight can be broken down into these colors using a prism. Voilà: we understand color!

Well, maybe. If light is transportation to our cosmic destinations, then color is the language necessary to understand them. How do we interpret information that we cannot physically detect? One way is that we assign colors we can see to the types of light that we cannot see.

Before we go any further, we should state that all of the images of space in this book are completely real (unless specifically labeled as an artist's illustration.) This needs to be said because there is sometimes confusion about the new digital reality. We live in an age where "to Photoshop" is a common verb, and there are countless ways to alter an image. If you can't trust that the photo of a celebrity on the cover of a magazine hasn't been sliced and diced, then how can you trust that these spectacular images from space are actually real?

We can understand the possible confusion and skepticism. But we can also say with confidence that the images on these pages represent real data of real things in the real Universe. Putting them together can take a lot of steps, but we want to go over the basics so that you can trust what your eyes will see.

In years past, astronomers used to take photographs of the sky using film. They would assign colors to different filters (or slices) of light from the telescope and stack them together. After some struggle in a darkroom, a color image of the heavens would emerge.

Today, virtually all data from telescopes are digital. In order to make these images, colors are assigned to different slices of data. In some cases, it will be three different slices of the same kind of light, such as optical light (that is, what we can see with our eyes). In other cases, it could be a layer of infrared data stacked on top of radio waves on top of X-rays, all of the same field of view. Any configuration of layers is possible if the data are available. So to really understand what is going on in any astronomical image, it's important to read the captions.

We are ever more accustomed to seeing scientific imagery in different types of light on our local weather forecast, even if we do not perceive it as such. Bright red patches on a satellite image may warn us of areas that can expect severe thunderstorms, for example, and green or blue patches might represent areas that can expect less severe storms. Since we are familiar with the concepts of weather, this likely doesn't intimidate us or put us off. We know that the thundercloud above us is not really bright red in the sky, but the color on the forecast image provides us with more information than just showing a solid dark gray cloud over a large region.

We also know that though we can't see an incoming cold front with our eyes, we should still bring a sweater for the day. The same type of thinking can be extended to astronomical objects and the images we are showing you from space. The phenomena might be more foreign, but the images represent actual data that have been colored to make them informational and accessible.

NOT BY THE NUMBERS: COLORING SPACE IMAGES

This might be a good time to look at "false color." Some people believe that space images are not real if the colors are randomly assigned to the data. This is not the case. Say you have a basic wool dress, and you dye it pink. What if you chose to dye it blue instead? Does that make one real and one fake? The two dresses are equally legitimate—they are simply different colors. A color choice doesn't change the structure or the fabric of the dress; you just gave it a color. The same can be said of these images from space. The color is a choice made by the people who put the images together, usually to help show specific phenomena or to make the meaning of an image clearer, but the data that are being represented are still real.

What might the local news show as your nightly weather forecast? A satellite image in infrared light so that the cloud cover is visible.

Let's take a moment to look at the whole Earth in a various kinds of light and using a range of colors.

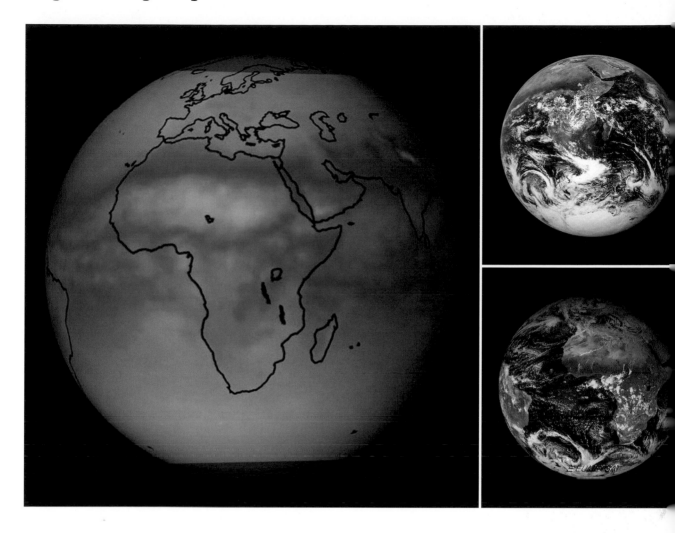

These three images show the whole globe of Earth in different types of light—(clockwise) ultraviolet, optical, and infrared—as seen from satellites.

What you see in this panel of images is Earth in different types of light. It is still the same planet we have always known and loved. The images just show that different features are emphasized in the different kinds of light.

Let's keep this example of the multicolored Earth in mind as we travel farther away from our home planet. You may be less familiar with these cosmic objects, but they are being shown in their true state, even if the image comes in different colors and types of light.

It's time to get off our planet now. Don't worry if the concepts we have put in your cosmic travel kit about the nature of light and color don't all make sense yet. One of the purposes of traveling is to learn, and there is no better classroom than the Universe we all live in.

EARTH

The Earth. Terra Firma. Gaia. We use all of these names for our home in space. It took astronauts and spacecraft that could get a good distance away from Earth before we could capture such a portrait of our home planet. This updated photo of our planet comes from NASA's most recent "Blue Marble" series of observations and was taken in January 2012.

U.S. EAST COAST AT NIGHT

Here is what part of the East Coast of the U.S. looks like at night when seen from the International Space Station. Starting from the bottom-left and moving to the right, regions in this densely populated area include Hampton and Richmond, Virginia; Washington, D.C.; Baltimore, Maryland; Philadelphia, Pennsylvania; and New York City and Long Island, New York (those bright lights on the lower right). Bits of Connecticut, Rhode Island, and Massachusetts can be seen above Long Island.

EARTH AT NIGHT

This is a series of images composited together to show what the Earth looks like at night. (This could never be just one snapshot, since half of the Earth is in daylight at any given time.) The point of this image is to demonstrate how much light escapes up into space from human development. Not surprisingly, most of the light comes from areas with the densest populations. If we ever did get visitors from somewhere else, it wouldn't be hard to figure out where most of us live.

HURRICANE ISABEL FROM SPACE

Hurricanes can be devastating events to witness and experience from the ground. From the tranquility of space, however, they can be quite beautiful. This photograph, taken from the International Space Station, shows Hurricane Isabel as it headed for the East Coast of the United States in September 2003. Photographs of hurricanes viewed from above can provide details on the storm's structure that scientists can use to better understand dynamics of the storm.

THUNDERSTORMS FROM SPACE

Thunderstorms and circular cloud patterns were photographed by International Space Station (ISS) crewmembers as they looked out at the blue line of Earth's horizon on October 6, 2009. The astronauts were orbiting near the Rio Madeira near Bolivia. The striking patterns of these clouds are probably caused by aging thunderstorms. The reflection of sunlight off the waters of the Amazon Basin back to the camera onboard the ISS can be seen, most noticeably toward the lower right.

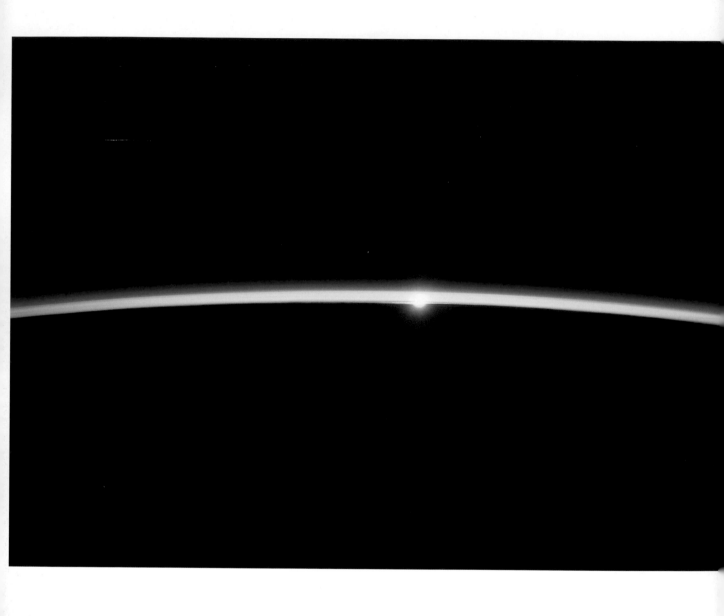

THE THIN BLUE LINE

Our planet's thin atmosphere is all that stands between life on Earth and the cold, dark void of space. Earth's atmosphere has no clearly defined upper boundary but gradually thins out until it is no more. The layers of the atmosphere have different characteristics such as protective ozone in the so-called stratosphere, and weather in the lower layer. This photograph, which was taken by the crew of the International Space Station in 2008, also shows the setting Sun as it crosses our all-important but relatively thin atmosphere.

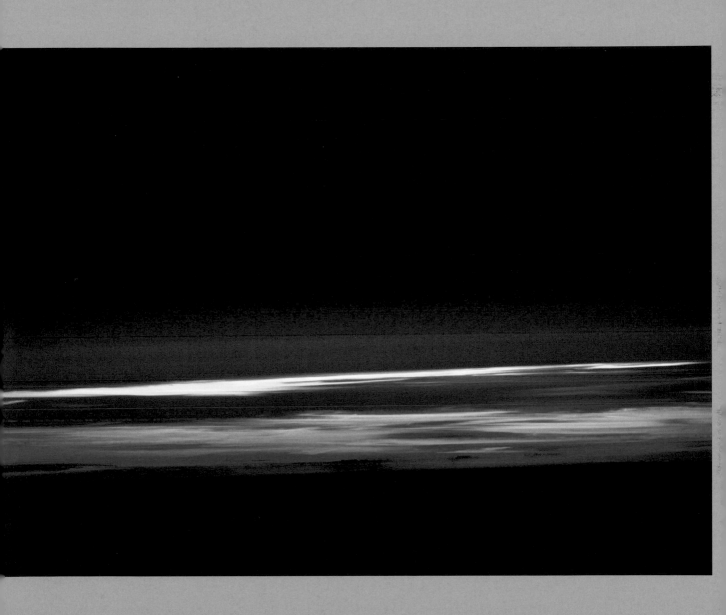

SUNSET FOR EARTH'S HORIZON LINE

This photograph from astronauts on board the orbiting International Space Station (ISS) shows the Earth's limb, or horizon line, at sunset. The bottom of the image contains the part of Earth that was already experiencing night, with the blackness of outer space draped over the very top. In between the two black areas are the troposphere (where most clouds live), stratosphere (where airplanes tend to fly), and the very highest parts of the atmosphere that gradually fade into the vacuum that is outer space. Sunsets are relatively common sights for astronauts on board the ISS—they can be seen up to sixteen times each day.

GLOWING LINES ACROSS THE SKY

Auroras are one of the best shows in the night sky when seen from the ground, and they are even more impressive when viewed from above. High above our planet, astronauts on the International Space Station get up-close views of auroras from their windows as the space station flies through space. NASA astronaut Doug Wheelock took this photograph on August 13th, 2010, when a minor solar wind stream hit Earth's magnetic field. The impact was not strong enough to generate auroras that were visible for Earthbound folks.

BIRDS-EYE VIEW OF A COMET

Some comets—known as "sungrazers"—pass extremely close to the Sun. Many of these comets evaporate, but some of the larger ones survive for a return flight back to the outer Solar System. Comet Lovejoy, shown in this picture, was one of the comets able to emerge from its close encounter with the Sun. This photograph shows Comet Lovejoy as it appears to be plunging into Earth. In reality, the comet is traveling safely far away from our planet.

3
OUR MOON AND SUN

Three things cannot be long hidden: the Sun, the Moon, and the truth.

— GAUTAMA SIDDHARTHA

ALL RIGHT, COSMIC EXPLORER, WE'RE JUST ABOUT READY to push off for our first destination, the Moon.

Our Moon is about a quarter the size of Earth and very close to us, astronomically speaking. It has important effects on the Earth, including helping to produce tides in our oceans through gravity. At a distance of about 240,000 miles (386,000 kilometers) from our planet, it is by far the nearest neighbor we have. That's why the Moon is the only celestial body we have yet been able to send astronauts to, as we did with the Apollo missions back in the 1960s and 1970s.

Perhaps the most attention-grabbing aspect of the Moon from a casual skywatcher's perspective is that it appears to change from night to night in the sky. In reality, the Moon does not alter its shape over the course of a month. Instead, the Moon's position in the sky changes as it rotates around the Earth every twenty-seven days or so. The different points in its orbit allow more or less of the Moon's disk, or face, to be illuminated by the Sun at any given time. This is what gives those of us on Earth a different view of the Moon as the month goes by.

Each of Moon's monthly phases has a name. When the Moon is directly between the Earth and the Sun, we call it the "new moon." We cannot see it then. When it is on the other side of the Earth, it is the familiar "full moon," its disk fully lit up by the Sun. In between those two phases, the Moon is either showing us a little more ("waxing") or a little less ("waning") of itself depending on how much sunlight is able to reach it.

To see this a little more clearly, take a flashlight and place it on a table or other solid surface in a dark room. Then put a tennis ball or a table tennis ball on a stick and hold it at arm's length. Standing still, move the ball around your body. Notice that how much of the ball is illuminated depends on where it is in relationship to the flashlight. This is a very rough approximation of what happens during our Moon's journey around the Earth.

PREVIOUS SPREAD:
Above the dark surface of the Earth, an orange-red glow and brown edge reveals the lowest and densest layer of atmosphere called the troposphere. The faint white-gray layer above that is a slice of the so-called stratosphere. The upper parts of the atmosphere—the mesosphere, thermosphere, and exosphere—fade from blue to the blackness of space.

OPPOSITE:
Photographed during the Apollo 10 mission in May 1969, this image shows a number of large craters on the moon.

FORMATION OF THE SOLAR SYSTEM

One intriguing aspect of the Earth-Moon relationship is that the Moon orbits us in such a way that it always shows the same side to us. This has led to the expression "the dark side of the Moon," meaning a place that is secret or unknowable. While the dark side of the Moon has intrigued skeptics and science fiction buffs for centuries, the surface that we can see from Earth is just as captivating. Even without a telescope, we can see large swaths of dark and light areas on the lunar surface. You may have noticed that the Moon looks pockmarked. Those overlapping circles of different sizes and colors are craters. But where do those craters come from?

Illustration of the formation of the solar system.

The answer goes back to how the Solar System formed. Billions of years ago, the first inkling of our Sun was sparked when a massive cloud of gas and dust collapsed. Over time, this central core of the cloud gathered enough mass for the nuclear fusion—the process that powers the stars—to ignite.

After the baby Sun got going, there was still quite a bit of material swirling around it. This material was flattened into a rotating disk that circles the Sun. As time moved on, clumps of material in this disk began to collide into each other. The bigger the clumps grew, the more material in the disk was drawn to them through their increased gravitational prowess. Think of a ball of pie dough that rolls around on the cutting board, gathering the smaller pieces of dough and flour as it moves.

After a while, a handful of clumps gained enough material that they started to dominate the disk. These "superclumps" eventually became what we call the planets. While the forming planets pulled in most of the material in that circling disk, many smaller chunks of rocks remained floating around the infant Solar System. These orphan rocks would sometimes slam into the very young planets, causing impacts from relatively small dings to enormous wallops.

All of the planets were bombarded like this during their early years. On Earth, however, with its atmosphere, cycles of weather, active volcanoes, and vegetation, most of the evidence of these impacts has been erased from the planet's surface. But sometimes that evidence still exists right out in the open, as at Arizona's Meteor Crater, which is an excellent example of an impact crater that survived for billions of years.

However, for planets and bodies that had neither dynamic weather nor active geology to disguise that evidence, there was nothing to cover the tracks of the early impacts, so they remain visible today. The Moon, as well as Mercury and other bodies in the Solar System that we will talk more about in the next chapter, are current visual reminders of the violent history that all of the bodies in the Solar System share.

Meteor Crater, also known as Barringer Crater, in Arizona, is about 50,000 years old and 550 feet (150 meters) deep.

Today, the Moon is without life or an atmosphere. That doesn't mean that it's not interesting to both scientists and would-be explorers. The Moon contains vast stores of minerals, and perhaps significant amounts of water deep within its layers. We have only scratched its surface, quite literally.

Depending on how the political winds blow in the future, government agencies may or may not be sending more people back to the Moon in the next decade or so. Perhaps, because of those minerals or that water, there will an incentive big enough to get the private sector motivated to travel there. One thing remains certain: the Moon has long been a source of fascination to us and will continue to be for those of us who like to dream about it.

THE HUNT FOR EXO-PLANETS

People sometimes argue about whether there are eight or nine planets in our Solar System, depending on if we count Pluto. This might soon be a moot point, however. That's because the number of planets we are finding outside the Solar System continues to rise—and fast.

At last count, there were hundreds of confirmed planets outside of our Solar System, with thousands more labeled as "candidates" (which means that scientists need more data to confirm what they are.) The first of these "extrasolar planets" or "exoplanets" was discovered in 1995, and the race to find more and different kinds of planets has been in a full sprint ever since. So far, many of the extrasolar planets found have been giants the size of Jupiter or even much larger. This does not necessarily mean that big planets dominate the Universe—only that our technology more easily allows us to find big planets at this point.

Astronomers have used a clever technique that does not directly observe these planets. Instead, they look at the star that a planet (or planets) orbits around and search for tiny tugs that the star experiences because of gravity. It can take years of steady observations and careful analysis to find these minute effects, but the technique has proven to be very effective. However, it is biased toward finding only the largest of these extrasolar planets since they have the biggest gravitational effects, relatively speaking, and give astronomers a fighting chance to find their signals in all of the data.

In the past several years, another reliable tactic has emerged for astronomers to look for worlds outside of our Solar System. A "transit" refers to when one astronomical object passes in front of another (you may have heard of Venus's transit of the Sun that occurred in June 2012). In the case of planet hunting, this means that astronomers can watch for minuscule dimming

OUR SUN

The next stop on our trip is that other all-important fixture in our sky: the Sun. In the realm of space, one of the most overlooked facts may be that our Sun is, in fact, a star. We have a particular affinity for the Sun—and for good reason. The glowing orb that rises and sets in the sky every day is responsible for our very existence. After all, if we did not have the heat and light from the Sun, we would not be able to live on Earth.

We may be a bit complacent, though, when it comes to appreciating just what a magnificent object our local star is. From the surface of the Earth, the Sun looks like a bland yellow disk.

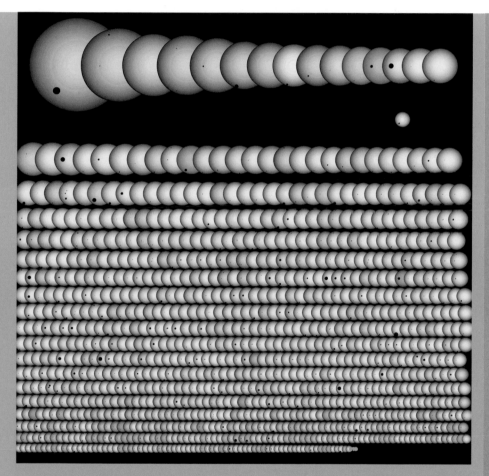

This illustration shows the collection of the Kepler mission's planet candidates (black dots) with their parent stars. The parent stars are arranged in order of their size from the largest (top left) to the smallest (bottom right). Some stars are shown with more than one planet in transit with them. For reference, our Sun is shown at the same scale as the rest of the parent stars; see the lone star just below the top row on the upper right. In silhouette against the Sun, Jupiter and Earth are shown.

of light from stars as a planet orbits in front of it. NASA's Kepler Mission, scheduled to continue operations until 2016, has been making new planet discoveries left and right using this transit technique.

While we have not found a planet exactly like Earth yet, the hunt is on. Many scientists think that we may not have to wait too much longer to find one.

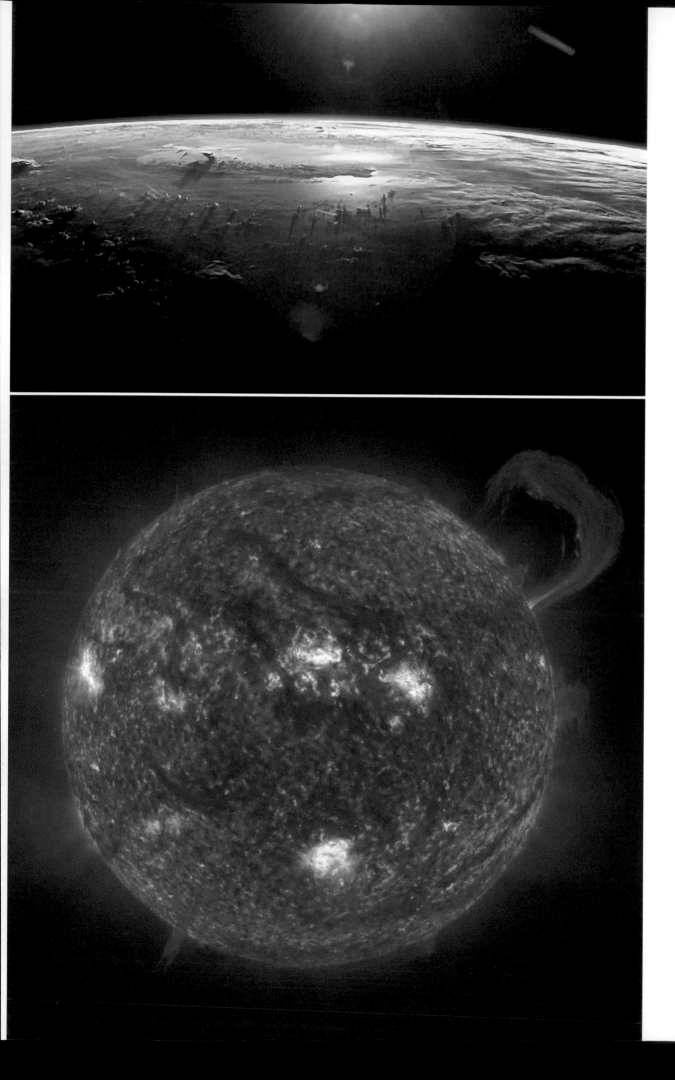

Modern telescopes, however, have given us great visual proof that the Sun is anything but calm. In fact, the Sun is a dynamic tempest of superheated gas that radiates in ways that we could never see just using our eyes.

OUR SUN IN A WHOLE NEW LIGHT

Here our conversation about color and light in the previous chapter begins to become relevant. Remember that a single object emits many different kinds of light. Let's look at our star, the Sun, in radio, infrared, ultraviolet, and X-ray light.

Our Sun is the brightest source of radio waves in our sky. The most active regions in these images (below) are the brighter and whiter areas. The infrared image shows darker regions where the gas is cooler and denser. In the lower right area of this image of the Sun, you can see a small eruption of gas where the temperatures reach millions of degrees. The ultraviolet image shows more of the higher energy flares and coronal mass ejections. In the X-ray image, there are coronal loops and streamers. Darker areas are cooler regions where the gas is less active. These are all images of the Sun; they just show different features depending on what type of light is being observed.

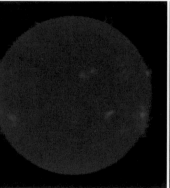
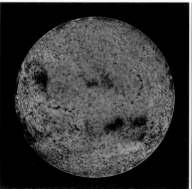
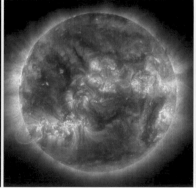
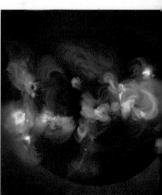

A WEATHER CHANNEL FOR THE SUN

You might be surprised to learn that the Sun also experiences weather. Solar weather, or space weather, is not like Earth's— that is, there's no rain, snow, or sleet. Rather, the weather on the Sun comes from the dance and play between huge tentacles

The Sun as it appears in radio, infrared, ultraviolet, and X-ray light.

of hot plasma and magnetic fields that stretch out from the solar surface. This results in enormous storms of charged particles that erupt from the Sun. The effects of these solar storms can reach for millions of miles, including out to the Earth and its atmosphere. If one of these storms is powerful enough, it can knock out satellite transmissions and produce spectacular sky light shows called the aurora, also known as the Northern Lights in the Northern Hemisphere and the Southern Lights in the Southern Hemisphere.

A close-up view of activity on the surface of our Sun.

Some solar storms cause the aurora, called the Northern Lights in our hemisphere.

AS THE SUN CHANGES: THE SOLAR MAXIMUM

Our Sun is a changing sphere. There is constant movement on the Sun, and storms on its surface affect regions millions of miles away in space.

In addition to this daily changing activity, the Sun is also on an even bigger cycle—in fact, a cycle that lasts about eleven years. During this cycle the Sun's magnetic poles flip: North becomes south, and vice versa. This pole flipping causes the activity on the Sun—including the number of storms and ejections in space—to increase. Scientists predict that the "solar maximum" in the most recent cycle will occur in late 2013 or 2014. During this upcoming solar maximum, your cellphone or satellite TV could experience more glitches or blackouts during that time. If you call for technical support when that happens, your service provider might suggest you take it up with the Sun.

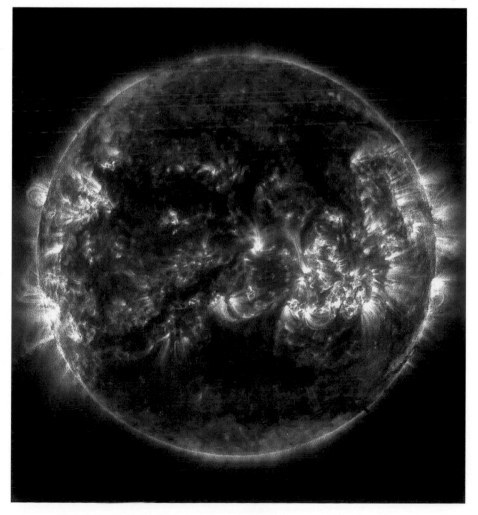

An ultraviolet image of the Sun during the last solar maximum, taken on February 22, 2002.

SOME
SUN AND
MOON
FACTS

OUR POCKMARKED MOON IS A VISUAL REMINDER OF THE VIOLENT
HISTORY THAT ALL OF THE BODIES IN THE SOLAR SYSTEM SHARE.

OUR NEAREST STAR, THE SUN, IS THE LARGEST OBJECT IN OUR
SOLAR SYSTEM.

THE SUN IS DYNAMIC AND TEMPESTUOUS.

DISTANCES ON A COSMIC SCALE

As the Earth orbits around the Sun, it keeps an average distance of about 93 million miles (150 million kilometers) between it and our star. The Earth completes one of these trips around the Sun once every 365.25 days—in other words, once a year. (That little one-quarter of a day left over is why every four years we add a day to the end of February, leap day, to keep the books balanced.)

Since we brought up the subject, now is a good time to start delving into the hard-to-fathom idea of cosmic distance. When it comes to space, the distances we encounter can be unimaginably big and brain-achingly vast. So far, we have tackled our two most famous cosmic neighbors: the Moon and Sun. At their respective distances in the hundreds of thousands, and millions of miles (or kilometers) from us, they sound very far away.

In some ways, they are. But in others, they aren't even out of our celestial backyard. Things in the Universe are so far apart that most of the time astronomers typically abandon the units of distance we use on Earth—miles, kilometers—and use their own.

One unit that astronomers often employ, and that we will use in this book, is the light-year. From the sound of it, this should be a unit of time. Instead, it is a unit of distance. A light-year is the distance that light will travel in one year, which translates into about six trillion miles. Six trillion is a six followed by twelve zeros!

How does that work, exactly? Let's pretend that you can walk exactly one mile in fifteen minutes—no more and no less. If you were to say that you had been walking for fifteen minutes, then that can be translated automatically into one mile of walking.

A thirty-minute walk is equal to two miles, and so forth. In this way, the unit of time (minutes) becomes a unit of distance (miles) because the speed of the walking in this hypothetical situation never changes.

This is exactly how the light-year works, because speed of light is absolutely constant, as we mentioned in the previous chapter. In fact, this consistency of light's speed is one of the most

Our Earth is the bright white dot in the lower left of this image, with the Moon seen as the smaller white speck just to the right of Earth. When NASA's MESSENGER (it's written in all capital letters because it's an acronym for Mercury Surface Environment, Geochemistry and Ranging) mission took this photograph in May 2010, there were 114 million miles between Earth and the spacecraft. To compare, 93 million miles separate the Earth and our Sun on average.

fundamental ideas in all of physics. (You can thank Einstein for that one, too.) It has been tested and tested over the years, and that simple fact that light travels at a constant speed has always stood up. If you want to do the math, light moves at just about 186,200 miles every second. That's fast. The fastest land animal on the planet, the cheetah, can reach speeds of only seventy miles an hour.

As we explore objects beyond the Moon and Sun, it will quickly become obvious why astronomers devised this different way of keeping track of distance. For example, the nearest star beyond our Sun, Proxima Centauri, is about four light-years away. That's somewhat manageable to talk about in terms of miles: about 24,000,000,000,000 (that is, 24 trillion miles, or 38 trillion kilometers). But if you consider something still relatively nearby, such as the center of the Milky Way, you are looking at a distance of about 26,000 light-years away from Earth. That value translates to 156,000,000,000,000,000 miles, or 250,000,000,000,000,000 kilometers. That's a lot of zeros to keep track of.

So you can either think of the distances to the Moon and Sun in our familiar units of miles or kilometers. Or you can try to get more comfortable with the navigation of space, which translates to the Moon's being about 1.3 light-seconds (how far light travels in 1.3 seconds) and the Sun just a hair under 500 light-seconds away, or about 8.3 light-minutes. For purposes of this book, we will give all of the distances to objects in space in terms of light travel time, that is, in the units of light-seconds, light-minutes, or light-years. Look for these reference points next to the titles of the objects in the image galleries.

As we will see in the next chapter, once we jump out to other objects in the Solar System and beyond, having the concept of "light-years" in our cosmic toolkit will come in very handy. Please note that for objects in the Solar System, in this book we use the average distance in light-travel time from the Sun (instead of from the Earth). We do this because the planets in the Solar System—including Earth—travel in highly elliptical orbits and at different speeds around the Sun, so the distance between Earth and other planets can vary wildly depending on where each body is in its orbit. The Sun, however, makes a better starting point to mark off distance for planets and other local bodies in the Solar System.

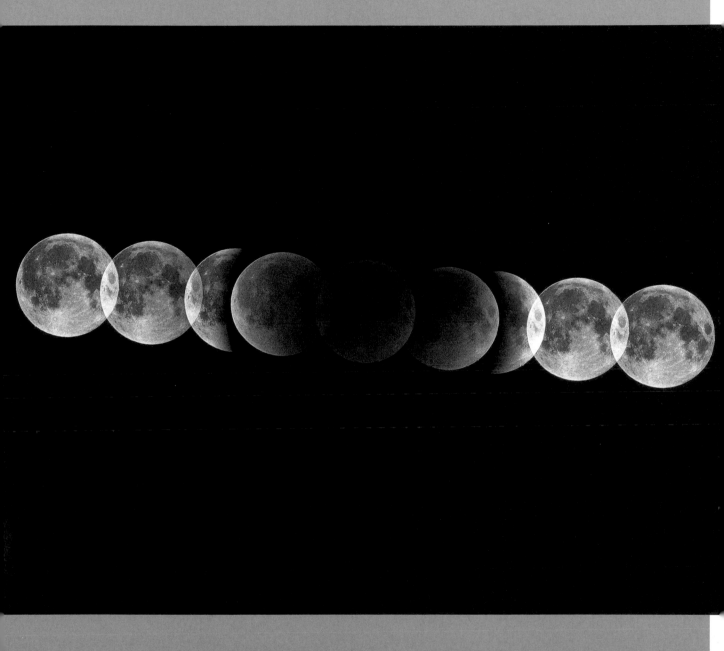

A LUNAR ECLIPSE: 1.3 LIGHT-SECONDS

When the Moon passes through the shadow of Earth, as this sequence of time-lapse images shows, it is called a lunar eclipse. During the lunar eclipse, light—mostly red light—is bent by the Earth's atmosphere, allowing only the red part of sunlight to reach the Moon. This filtered light is in turn reflected to Earth, causing the Moon to appear red.

FULL MOON: 1.3 LIGHT-SECONDS

A familiar sight to us all, the full Moon graces our night sky every month. The lunar landscape is a mixture of bright highlands and dark "seas" once filled with lava, both of which now show the scars of large impact craters and rays of ejected material. Scientists think the Moon itself was formed after a violent asteroid collision with the Earth billions of years ago.

FAR SIDE OF THE MOON: 1.3 LIGHT-SECONDS

Because the same side of the Moon always faces Earth, humans had to wait until 1959 for a Russian spacecraft to capture the first photographs of the far (or "dark") side of the moon. This image of the Moon's far side was composited from more than 15,000 separate photos taken by NASA's Lunar Reconnaissance Orbiter Camera between November 2009 and February 2011.

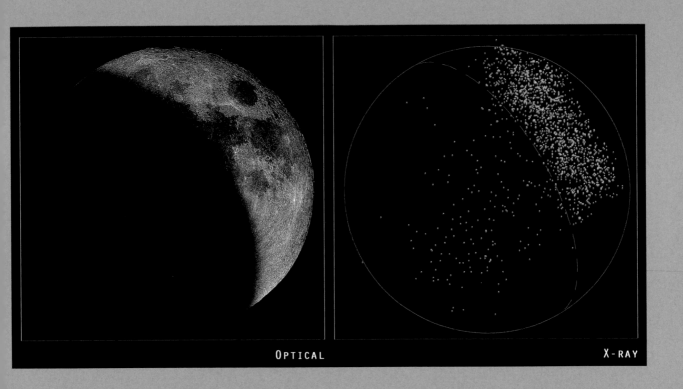

Optical

X-RAY

THE MOON IN A NEW KIND OF LIGHT

As you may recall, X-rays are supposed to come from things very hot or energetic. The Moon is cold, though, and doesn't give off any heat. So why would astronomers want to take an X-ray picture of the Moon? Because when the light from the Sun bounces off the Moon, it lets us see what is on the lunar surface. This X-ray image (right) from NASA's Chandra X-ray Observatory of the bright portion of the Moon reveals oxygen, magnesium, aluminum, and silicon atoms that are produced when solar X-rays bombard the Moon's surface.

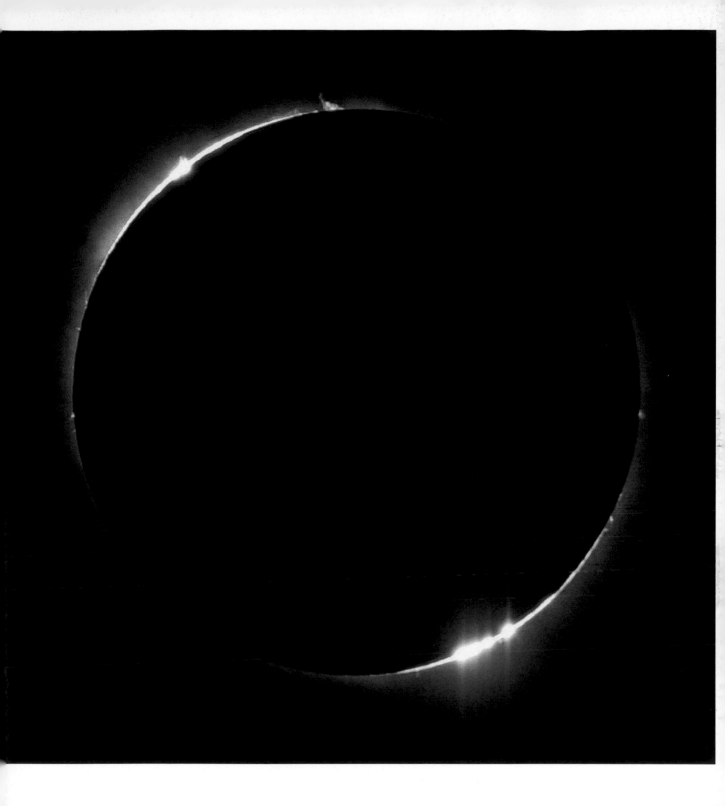

TOTAL SOLAR ECLIPSE: 1.3 LIGHT-SECONDS (MOON) AND 8.3 LIGHT-MINUTES (SUN)

A total solar eclipse is a natural phenomenon that happens when the Moon passes between the Sun and Earth in just the right alignment, blocking the light from the Sun to certain places on Earth. Some people, often called "eclipse chasers," will travel to virtually any spot on the globe to experience a solar eclipse. Totality, as shown in this montage of two separate photographs, occurs when the shadow of the Moon blocks the entire disk of the Sun, leaving only its outer layer (corona) visible. This solar eclipse was photographed from Turkey in March 2006.

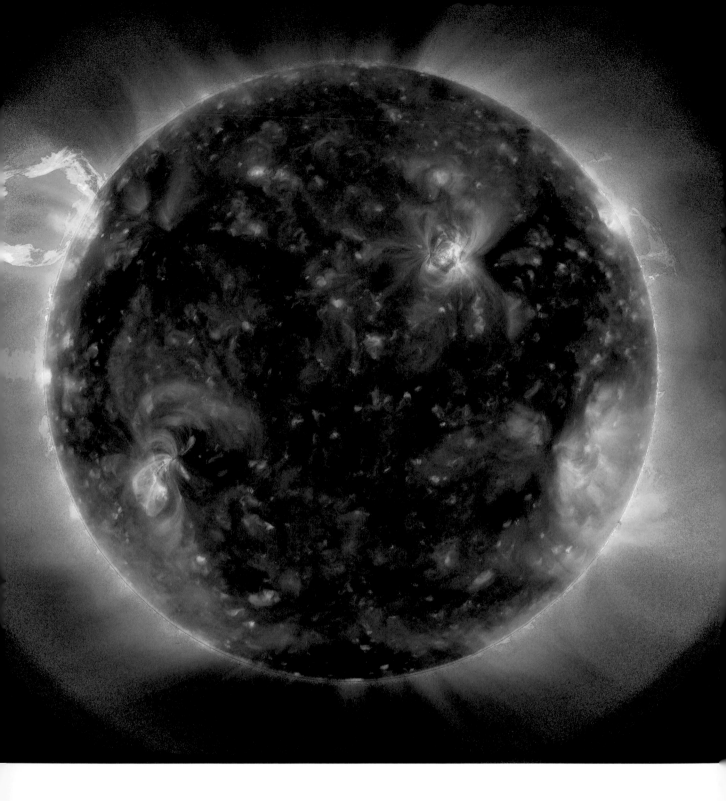

OUR SUN: 8.3 LIGHT-MINUTES

When the Sun is stormy, it results in streams of energized particles that can create auroras on Earth and can even disrupt cell phones. Taken in ultraviolet light, this image depicts the Sun's turbulent atmosphere, where the loops are constantly changing areas of energy and magnetism that drive what is known as "space weather."

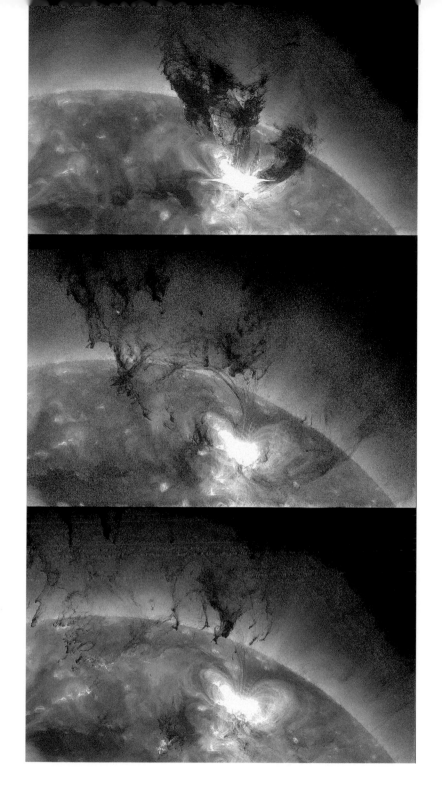

SOLAR FLARE: 8.3 LIGHT-MINUTES

This sequence of images of our Sun covers just thirty minutes of time on June 7, 2011. You can see a medium-sized solar flare (the bright white area in the top image) and a huge ejection of mass from the Sun (the darker material). A large glob of particles mushroomed up and then fell back down to the Sun. The very large eruption of gas looked as if it covered an area almost half of the Sun's surface. NASA's Solar Dynamics Observatory (SDO) recorded the images in the highest ranges of ultraviolet light.

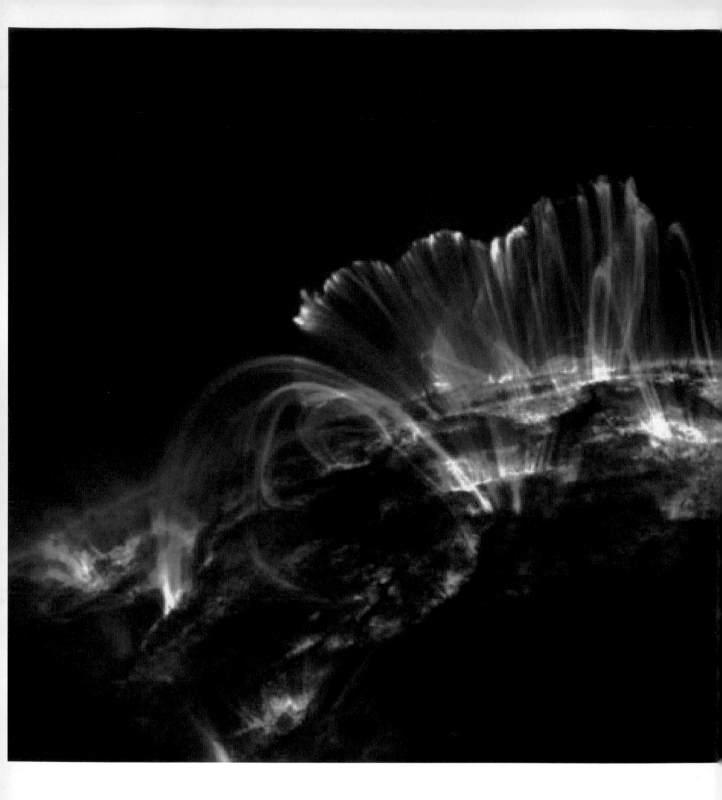

CORONAL LOOPS: 8.3 LIGHT-MINUTES

This close-up view of one of the many storms seen on the Sun is like zooming into one particular storm in North America, as compared to seeing the weather of the whole globe. The "average" storm on the Sun is huge and could easily engulf the entire Earth. These loops are both beautiful and dangerous. The structures are made of blisteringly hot, electrically charged gas in the Sun's atmosphere. Sometimes these loops snap and the hot gas falls back onto the Sun in what is known as "coronal rain."

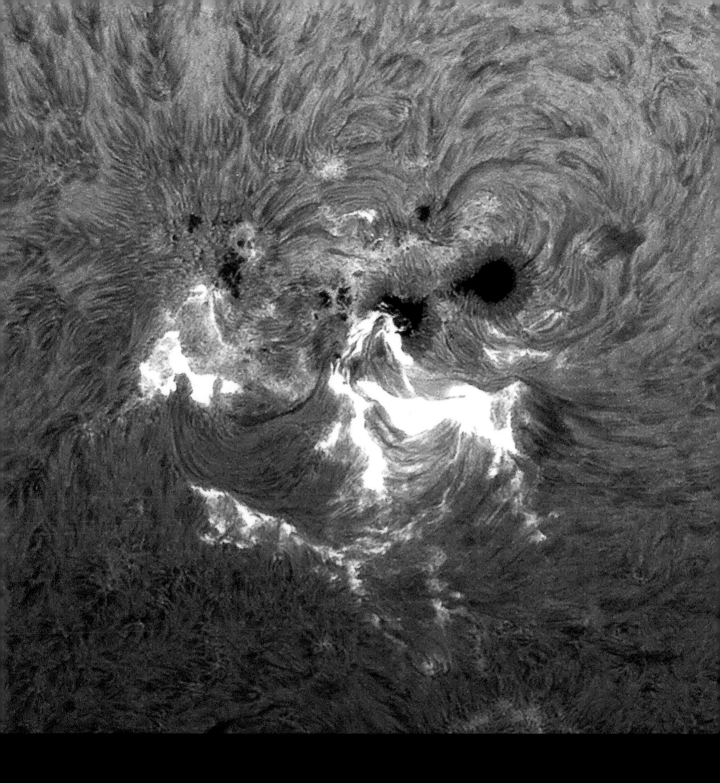

SUNSPOTS: 8.3 LIGHT-MINUTES

What are the dark spots on the Sun? They are sunspots—temporary regions of reduced surface temperature caused by increased magnetic activity. Solar flares, or outbursts, emanate from sunspots. The NASA/ESA Solar and Heliospheric Observatory (SOHO) mission recorded a giant eruption from this sunspot, throwing out energetic particles that hit the Earth about 48 hours later. Such events can disrupt satellite communications, but they are also responsible for the dramatic and beautiful phenomenon of auroras that some people at Earth's high northern and southern latitudes are lucky to observe.

SUN IN HYDROGEN-ALPHA: 8.3 LIGHT-MINUTES

Looking at the Sun in just the light of hydrogen atoms (a small portion of "visible" light), we see a dramatic view of our home star. Only eruptions from the surface of the Sun and arcs of gas are seen. The darker streaks are the plumes of gas viewed from above. At the bottom center of the image, small darker patches are visible. These are sunspots, depressions in the surface caused by the Sun's complex magnetic field.

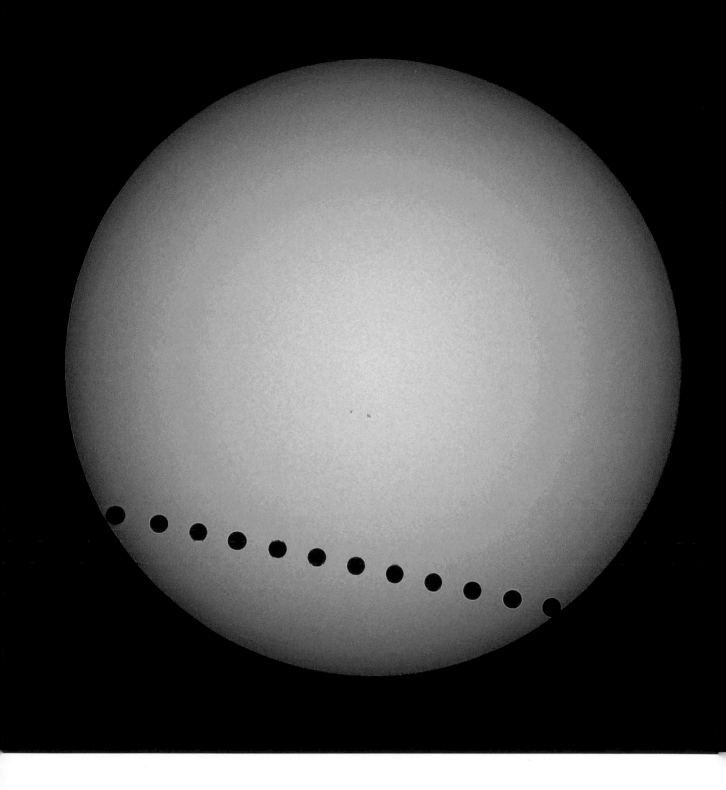

TRANSIT OF VENUS: 140 LIGHT-SECONDS (VENUS) AND 8.3 LIGHT-MINUTES (SUN)

A rare treat for astronomers, this multiple exposure shows the planet Venus in its stately five-hour transit as it passed between the Earth and the Sun in 2004. Transits occur in pairs eight years apart, but each pair occurs just once a lifetime. The most recent transits were in 2004 and in June 2012, for example. If you missed the last one, you will then have to wait until 2117 and 2125. Venus is most familiar to us as a bright point of light seen just after sunset or just before sunrise.

4

BODIES OF THE SOLAR SYSTEM

The solar system has no anxiety about its reputation.

— RALPH WALDO EMERSON

SO FAR, WE HAVE DROPPED IN TO SOME RELATIVELY familiar haunts: to push off for our first destination, the Moon. the Earth, Moon, and Sun. Now that we've checked out our local digs, it's time to venture a little farther out. There is a lot to explore in the Solar System, the equivalent of our cosmic neighborhood. There are some obvious places to stop along our walk down our cosmic block, including the planets: Mercury, Venus, Earth, Mars, Jupiter, Saturn, Neptune, Uranus, and Pluto—if you are counting Pluto as a planet, which we'll talk about later.

There are also other great attractions a little more off the beaten path across our Solar System. For example, most of the Solar System's planets have bodies, or moons, in orbit around them, most famously our Moon, as we discussed in the previous chapter. Having just one moon, though, may turn out to be the exception. Some planets, such as Jupiter and Saturn, actually have dozens of moons in orbit around them.

PREVIOUS SPREAD:

It might look like a familiar sight, but this sunset is not from our home planet. NASA's Mars rover Spirit captured this alien sunset in 2005 as it peered toward the western sky from its perch on Mars. The Mars Exploration Rovers have given us a perspective much like our own, but on another world millions of miles of away.

SUN · · · MERCURY VENUS · · EARTH · MARS · · JUPITER · · · SATURN · · · URANUS · · NEPTUNE PLUTO

COMET

This illustration shows the order of planets in our Solar System. They are not shown to scale, so that all the objects identified are visible and can fit on the page.

As we've discussed, the Moon—we use the capitalized version to mean the satellite of Earth, and the lowercase version "moon" to refer to the moons of other planets—is not terribly active these days. Some of the moons around other planets, though, are dynamic and evolving worlds. There are moons with liquid oceans of methane, others with frozen surfaces potentially covering bodies of water, and more. Think back on all those extreme environments we explored in the first chapter here on Earth. Very similar environments apparently exist on some of these moons, and they may end up being the best places to look for life off our home planet.

In addition to the planets and their moons, the Solar System has an asteroid belt between Mars and Jupiter that humans might be able to visit in the future. There are also interplanetary interlopers known as comets that sometimes make their way close to Earth on their journey around the Sun. In this chapter, we'll take a quick stop at a number of neighboring bodies in the Solar System and see what they have to offer.

THE INNER PLANETS

The Sun's position at the center of the Solar System dictates many things about the planets—largely because of how close or far away they are. The closest four planets, Mercury, Venus, Earth and Mars, are usually grouped together as the "inner planets."

MERCURY

Mercury is the closest planet to the Sun, at an average distance of just 36 million miles (58 million kilometers). This isn't a huge distance in astronomical terms. In fact, this close orbit means that Mercury whips around the Sun roughly once every eighty-eight days. In other words, a year on Mercury lasts less than three months here on Earth.

To the eye, Mercury looks a little like our own Moon—riddled with craters and sections of smoothness. While it's not obvious to the eye, Mercury is a planet of extremes. On the side facing the Sun, temperatures on the surface can soar to 800 degrees Fahrenheit (427 degrees Celsius). Considering that an oven broiler in your kitchen cooks at about 500 degrees Fahrenheit (260 degrees Celsius), that's hot.

The long streaks seen in this image of Mercury radiate from an impact crater shown at the very top of the image. The lines or "rays" from the impact crater extend for as much as 600 miles (1,000 kilometers). Such rays are formed when an impact excavates material from beneath the surface and throws it outward from the crater. Rays such as these fade over time as they are exposed to the harsh space environment. Craters with bright rays are probably young because the rays are still visible.

VENUS

Venus is the second planet from our Sun, and arguably it is Earth's closest counterpart in the Solar System. With a size and weight just a hair under Earth's own, Venus currently has active volcanoes, mountains, and rivers. In its past, Venus probably had liquid oceans of water.

However, today's Venus doesn't look like any place we'd want to live. Its thick atmosphere, mostly carbon dioxide, traps the heat the planet receives from the Sun. In some ways, it experiences the complete opposite problem that Mercury has in retaining heat. There is no argument over the impact this "greenhouse effect" has on Venus: It is incredibly hot on its surface, at a scorching 900 degrees Fahrenheit (482 degrees Celsius). This intense heat would have boiled off any Earthlike oceans long ago.

One unique quality of Venus is that it rotates counterclockwise. The other planets in our Solar System—including Earth—rotate clockwise. This means that someone standing on Venus would see the Sun rise in the west and set in the east. Why the strange spin? One idea is that it received a series of blows from large objects such as asteroids in its early days. One or more of these blows could have delivered enough of a punch to change the direction that the entire planet was spinning.

There is evidence that when it was young, Venus almost certainly had liquid water—an essential ingredient for life on Earth. Some scientists speculate that single-celled organisms could actually survive within the harsh yet stable atmosphere on Venus today.

EARTH

The next planet in the Solar System is our very own Earth. Although we've talked about Earth already, we should mention a couple of additional things that make Earth so special. As with all real estate, it's location, location, location. Earth happens to be in a sweet spot—not too far and not too close to the Sun. We have enough heat on our planet to have liquid water (very important to us humans), but not too much that we cook it right off the surface.

Combine this Goldilocks temperature with our thick atmosphere that contains good things like oxygen, and an active magnetic field that shields us from harmful radiation and debris, and the Earth turns out to be the choicest property in our Solar System.

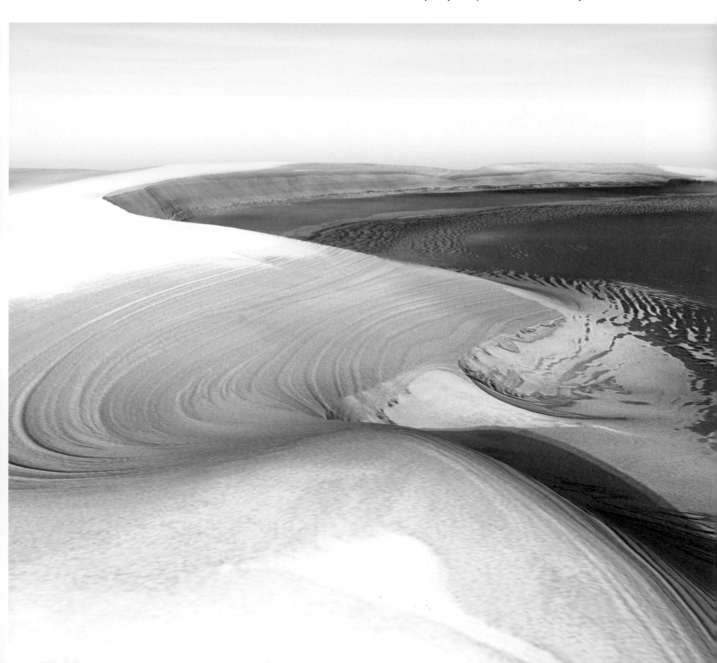

MARS

Mars is our neighbor on the other side of us, farther from the Sun. For decades, Mars has been a fixation of science fiction and actual scientific exploration. The Viking satellites that went to orbit Mars in the 1970s sent back much information on what we have long called the Red Planet, including some data that suggests there may have been liquid water on the planet at some point in its past.

NASA returned to Mars in spectacular fashion in early 2004 when the Mars Exploration Rovers, known as Spirit and Opportunity, safely landed on the planet's surface. These two plucky robots were supposed to last about six months. Instead, they both lasted over six years, and one was still returning science data as

In 2007, NASA's Phoenix mission landed in the northern arctic plains of Mars to study the history of water and potential habitability in the ice-rich soil. Phoenix verified the presence of water ice in the Martian subsurface, and found calcium carbonate, an indicator of a less acidic (more potentially habitable) planet in the past. Phoenix even observed snow falling from clouds in the Martian atmosphere.

of 2012. These rovers showed that the Viking results were just the tip of the iceberg. Spirit and Opportunity have found much more evidence for water, discovered key minerals in Martian rocks and dirt, and generally given us a never-before-seen view of the Red Planet's surface. A recent development in the exploration of Mars was the arrival of the Mars Science Laboratory in August 2012. This spacecraft carries the Curiosity rover, which is about the size of a Mini Cooper. By comparison, Spirit and Opportunity are each about the size of a golf cart, which means that Curiosity can carry more scientific instruments and should be able to travel much farther than its successful roving predecessors.

THE ASTEROID BELT

The term "asteroid belt" may conjure up a scene from *Star Wars* as a good place to hide from the bad guys, or, for gaming fanatics, a videogame from the 1980s. Whatever the association, we have a genuine asteroid belt in our own Solar System.

The main swath of asteroids is found in between Mars and Jupiter. About half of the mass in the asteroid belt comes from four big asteroids that are each more than 200 miles (322 kilometers) wide. The rest is made up of smaller rocks of various sizes. The existence of the asteroid belt dates back to the formation of the Solar System itself. A planet was likely trying to form in this slot, but the gravitational tug-of-war between Mars and Jupiter would not let it happen. Instead, the area became a sort of rocky planetary

This illustration shows a narrow asteroid belt filled with rocks and dusty debris that orbits a star similar to our Sun.

graveyard. Because it contains material from when the Solar System was very young, scientists are eager to study these asteroids as important relics of our cosmic history. Plans are being formulated to send probes (and one day, perhaps, humans) to explore these asteroids for minerals and other important resources.

THE OUTER PLANETS: GAS GIANTS, ICE GIANTS

JUPITER

The next destination on our journey away from the Sun is Jupiter, the largest planet in our Solar System. Jupiter is so gigantic that it holds over twice the mass of all of the other planets combined.

It is hard for us Earthlings to appreciate just how big Jupiter is. Most of us have an appreciation of the scale of things when we have something in our own experiences to reference. For example, the Rocky Mountains seem enormous when compared to the Appalachians, but small if you travel to the Himalayas.

Keeping a sense of scale in the Universe is tough because we quickly lose our frame of reference. We might have some sense of the size of Earth because we have studied maps or flown over parts of it in an airplane. We may be able to use such information to get a sense of how big Jupiter is: more than one hundred Earths can fit inside Jupiter. If you have marbles, try fitting a hundred into a glass bowl. You'll find you need a pretty big bowl. Jupiter, in short, is a big, big planet.

This illustration shows the relative size of Earth in comparison with the gas giants Neptune, Uranus, Saturn, and Jupiter.

Jupiter is much different from any of the planets we've talked about so far, not just because of its size but also because of its composition. Unlike Earth, which is mostly solid (rocky crust, liquid middle, and iron core), Jupiter is almost entirely gas.

Sometimes scientists refer to Jupiter as a mini "solar system" because it has many moons orbiting it, has a large magnetic field, and produces more heat than it gets from the Sun.

One interesting fact about Jupiter is that it has a storm in its atmosphere that has been raging for at least four hundred years. It is known as the Great Red Spot, and its size (about three times that of Earth) and duration make us appreciate the storms here on Earth that mercifully last just a few days at most.

Jupiter, the most massive planet in our Solar System—with dozens of moons and an enormous magnetic field—resembles a star in composition, but one that did not grow big enough to ignite. The planet's swirling cloud stripes are punctuated by massive storms such as the Great Red Spot, visible at the middle-right of this image, which has raged for hundreds of years. The moon Europa casts a shadow on the middle-left.

SATURN

Saturn is perhaps the most recognizable of all of the outer planets, with its amazing ring system circling its waist. If you look at Saturn through a small backyard telescope, the planet seems like a cardboard cutout from a cereal box, with its cartoonish shape. Seeing Saturn and its rings through the eyes of a large professional telescope on the ground or in space, though, shows how the ringed planet is truly something to behold.

First observed by the Italian astronomer Galileo Galilei with a small telescope more than four hundred years ago, Saturn's rings themselves are a marvel. The rings extend away from Saturn for about 50,000 miles (80,000 kilometers). That's about six times the diameter of the entire Earth. For being so wide, though, Saturn's rings are incredibly thin, averaging only about 30 feet (10 meters) in height.

These rings are made up of countless small particles ranging in size from a dust grain up to a boulder. While it is still a question up for debate, many scientists think that the rings formed when a would-be moon collided with Saturn billions of years ago. The thinking is that this doomed moon was pulverized, and then its remains eventually fell into orbit around the center of the planet.

Saturn's rings cast a shadow back on the planet, as captured here by NASA's Cassini mission while the spacecraft was about 621,000 miles (about 1 million km) from Saturn.

LOOKING FOR LIFE

One of the most fundamental questions humans ask is: Are we alone? The jury is still out on whether other forms of life exist in our Solar System and beyond, but that doesn't mean scientists aren't trying to find more evidence. One of the longest running programs to look for life in space is the Search for Extraterrestrial Intelligence (SETI). There are a couple of different programs in SETI, but the major one involves using radio telescopes to pick up possible signals from other civilizations.

Antennas from the Very Large Array, located in New Mexico

Another way astronomers are searching for life is by sending spacecraft to places where they hope to find it within the Solar System. Scientists are sure that there are no advanced civilizations in our own Solar System, no matter what Hollywood would have you believe. Instead, scientists are looking for basic types of life such as microbes and other biological markers. Since Mars is very similar to Earth in many ways—including having liquid water on its surface in the past—there is a lot of speculation that it may have also harbored some sort of microbial life. So far, none of the instruments on the surface or in orbit around Mars has found any evidence for this, but astronomers will keep trying.

Scientists are also looking elsewhere in the Solar System, including important spots on some of Jupiter's and Saturn's moons. This may include under the icy surface of Jupiter's moon Europa, or perhaps in the liquid methane oceans of Titan, a moon of Saturn. Sending sophisticated instruments to these places, however, is very expensive and will take quite a long time. In the meantime, "astrobiologists" (scientists who study life and its evolution both here on Earth and beyond) are using all of the available telescopes already out there to learn as much as they can about these and other possible destinations for life in the Solar System.

URANUS

Even though astronomers discovered Uranus about 250 years ago, the planet remains somewhat of a mystery. Uranus was investigated in detail only once, when NASA's Voyager spacecraft flew by in 1986. Aside from that, scientists have had to rely on learning what they can through telescopes here on Earth and others in space, such as the Hubble Space Telescope.

There are some things that we do know about this planet. Uranus rotates on its side compared to the rest of the planets in our Solar System that spin like tops on the ground. This unusual orientation may mean that like Venus, which rotates backward, Uranus was the victim of a planetary hit-and-run by a large object early in our Solar System's development.

Uranus also has a lot in common with both its inner and outer neighbors, Saturn and Neptune. Like Saturn, Uranus has a ring system around its equator, though it is not nearly as large or spectacular as Saturn's. Uranus and Neptune are so similar that they are often referred together as the "ice giants." The name comes from the fact that they are quite cold owing to their large distance from the Sun—Uranus averages a distance of nearly 2 billion miles (3.2 billion kilometers) from the Sun, and Neptune orbits at about nearly 3 billion miles (4.8 billion kilometers) away from our star. For those of you who are keeping track in more cosmic terms, this translates into about 2.7 light-hours for Uranus and 4.2 light-hours for Neptune. And although Uranus and Neptune are made up mostly of hydrogen and helium (like Jupiter and Saturn), they also have relative high amounts of "ices" in their atmosphere. These ices are not all based on water, as we're used to here on Earth. Instead, some of these ices contain ammonia, methane, and other compounds that are deadly to humans.

Given its composition and far flung position away from the Sun, Uranus has the coldest atmosphere of any planet in the Solar System. Underneath that frozen cloud cover, scientists think, based on what they can determine so far, that Uranus has a solid core made up of ice and rock.

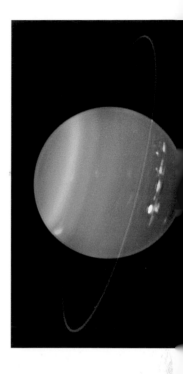

Uranus is the third largest planet in our Solar System. It has nine major rings and twenty-seven known moons. This image of Uranus, taken in infrared light, reveals cloud structures not normally visible to human eyes. Methane gas in the upper atmosphere absorbs red light, giving the planet its blue-green color.

NEPTUNE

Uranus's sister ice giant, Neptune, is also a fascinating world, but it is another one that we know relatively little about. No spacecraft has ever been sent there, and its super-dense atmosphere blocks the views of most telescopes. We do know that it has thirteen moons and a ring system, and that it takes almost 165 Earth years to orbit the Sun. Scientists have figured out that its thick atmosphere is very deep and gradually merges into water and other melted ices as you go farther down toward Neptune's surface. At the center of this murky cold ball is a solid core about the size of the Earth.

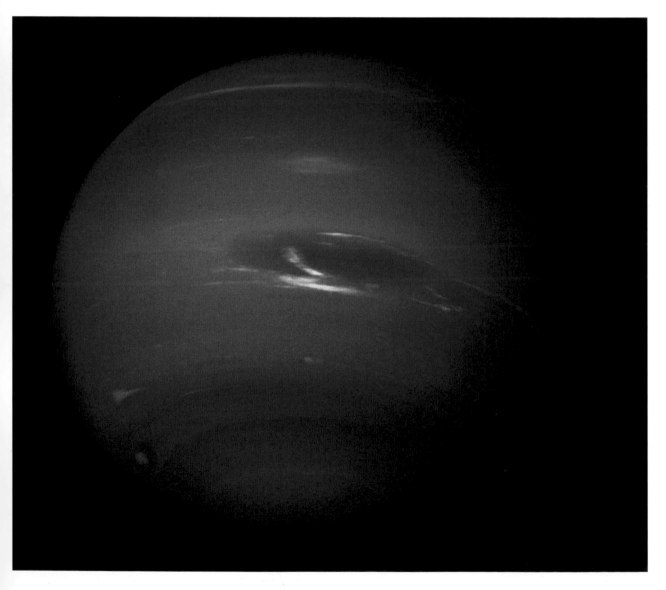

Dark, cold, and whipped by supersonic winds, Neptune is the farthest away from the Sun of all the hydrogen and helium gas giants in our Solar System. Its thick atmosphere acts as a dark veil, hiding the surface below.

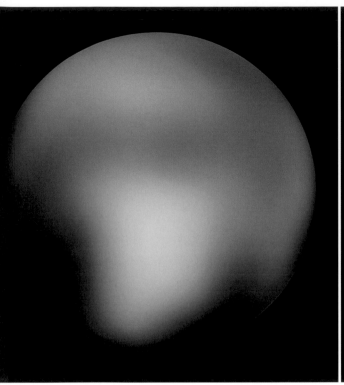

Pluto
(1400 miles)

Moon
(2100 miles)

Earth
(8000 miles)

PLUTO

We are now reaching perhaps the most controversial of all of the Solar System's objects: Pluto. Since it was discovered in 1930, Pluto has been described as the Solar System's most distant planet. That all changed back in 2006 when the organization responsible for classifying astronomical objects, known as the International Astronomical Union, decided to recast Pluto as a dwarf planet. For reference, Pluto is about one-third smaller than our Moon, with a diameter less than the distance between Boston and Houston (just under 1,500 miles, or 2,400 kilometers.)

Pluto's demotion didn't sit well with a lot of people, from scientists to schoolkids. There are some good reasons why this is all so hotly contested. At the root of the debate is the scientific process. We have learned a lot more about the Solar System since Pluto's discovery more than eighty years ago. Most relevant to Pluto is that we now know that the planets (whether you include Pluto or not) do not represent the end of the Solar System—far from it.

LEFT: *Astronomers have been able to detect changes on Pluto's surface by comparing Hubble images taken in 1994 with later images taken in 2002 and 2003 (shown here). The task is as challenging as trying to see the markings on a soccer ball forty miles away.*

RIGHT: *This illustration shows the relative size comparison of Earth, its Moon, and the dwarf planet Pluto.*

THE KUIPER BELT

Beyond Pluto's orbit, there is an ocean of small rocks and other bodies outside the boundary of where the planets orbit. This enormous rocky ring, containing what might be hundreds of thousands of objects, is known as the Kuiper Belt. The connection to Pluto is that the Kuiper Belt nudges right up to where Pluto sits. Astronomers have also found bodies in the Kuiper Belt that are even bigger than Pluto and just as close to the Sun as Pluto is. So this is the dilemma: If you call Pluto a planet, there is a strong case that you should call all or some of the bodies in the Kuiper Belt "planets" as well.

Whether you agree with this or not, Pluto is a very distant and cold world that astronomers are extremely curious about. In fact, NASA launched a spacecraft in 2006 called New Horizons. When it finally reaches Pluto in 2015, it should give us a better understanding of the special dwarf planet and its moons. Maybe then we can get some clarity on whether the decision to reclassify Pluto was the best one.

THE OORT CLOUD AND THE COMET BELT

Finally, in our tour of the Solar System, there is the Oort Cloud to consider. It lies even farther out in our Solar System—perhaps up to a light-year, or more than 6 trillion miles (9.6 trillion kilometers), from the Sun. We've never been able to take a picture of the Oort Cloud, since it's too far away from us and its objects are too small. What makes us think it exists then? Some comets have orbits that simply place them too far out to come from the Kuiper Belt, and there are other clues about the way comets travel that lead scientists to believe that a massive reservoir of rocks is out there. This and other circumstantial evidence suggests that the Oort Cloud is quite real and plays an important role in our Solar System.

Astronomers think that the Oort Cloud contains trillions of small bodies that generally stay in this very outer region of the Solar System. Occasionally, one of them gets nudged through a close brush with another and begins a long trek toward the Sun. One of the most famous of those visitors, Halley's Comet, is thought to come from the Oort Cloud.

Comets are one of the best ways for humans to be reminded of our unbreakable connection to space. For thousands of years, the appearance of a comet was considered a very bad omen. We now know that a comet doesn't mean imminent death or destruction. In fact, it can be considered good luck, since some scientists think that comets may have been responsible for actually delivering water to Earth during our planet's baby years in the very young Solar System. Today, scientists welcome comets because they bring far-flung bits of the primordial Solar System practically to our cosmic doorstep, which allows scientists to better investigate them.

Comet Machholz was discovered by Donald Machholz on August 27, 2004, and by January 2005 had become bright enough to be viewed from the Earth without a telescope. In this image, taken on January 7, 2005, Comet Machholz and its long extended tail are seen against the backdrop of the Pleiades star cluster.

Astronomers believe that comets come from these two distant structures of the Kuiper Belt and the Oort Cloud. The comets from the Kuiper Belt appear more frequently—that is, they show up every two hundred years or less. Comets that make their rounds less often are thought to originate in the Oort Cloud, which makes sense, since the Oort Cloud is very, very far away.

The comets themselves are bodies that contain dirt and ice, which is why they are sometimes called "dirty snowballs." As comets get closer to the center of the Solar System, radiation and wind from the Sun cause them to begin to light up. On the most basic level, comets have a "head" and a "tail." The head is a thin, fuzzy atmosphere that comes from the gases within the comet that come out as it begins to heat up. The tail comes from dust and gas being pushed back from the solar wind as it travels through the Solar System. Comets appear differently in our sky due to how close the comet gets to Earth as well as exactly how much gas escapes.

MOONS THAT ROCK

As we mentioned at the beginning of this chapter, there are many moons around the planets in our Solar System. Some of them are simply fascinating. Science fiction writers have been drawn to moons as intriguing destinations for a long time—think of Endor from *Return of the Jedi* or Pandora from *Avatar*. While none of the moons contains Ewoks or Na'vi, the moons in our Solar System offer compelling scientific reasons for exploring them. Here we describe just a few of our favorite moons.

EUROPA

The surface of Jupiter's moon Europa is a shell of ice about twelve miles deep. However, scientists have found evidence that beneath the ice Europa has an iron core, a rocky mantle, and an ocean of salty water about 60 miles (100 kilometers) thick. Jupiter's gravitational pull on Europa raises and lowers the sea beneath the ice, causing extreme tides. These in turn keep Europa's icy surface in motion, and they are probably the cause of the cracks and streaks seen in images of the moon. The same gravitational pull that affects the ocean tides is also thought to heat Europa's core, keeping the water from freezing entirely and driving geological activity. With both liquid water and a stable source of energy present, many astrobiologists and other scientists think Europa could be a habitat for extraterrestrial life.

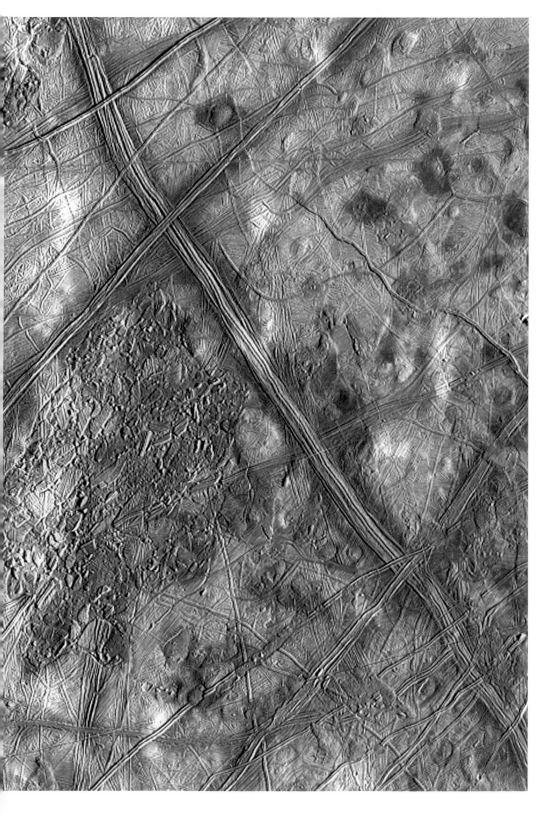

This image shows a close-up of Jupiter's icy moon Europa, which is about the size of Earth's own Moon. The dark brown areas in the image are thought to be salty material that comes up from geologic activity in Europa. The long, dark lines are fractures in its crust.

IO

About the size of Earth's Moon, Jupiter's moon Io is the most volcanically active body in the Solar System. Io's volcanoes erupt massive volumes of silicate lava, sulfur, and sulfur dioxide hundreds of miles or kilometers above the surface, constantly changing Io's appearance. The mixture of gases also means that Io is very stinky—probably something on the order of rotten eggs, though if we tried to find out by sniffing the toxic air, we would not live to tell the tale. Even though these volcanoes spew out different gases than ours here on Earth, some of Io's volcanoes have also been known to have lava erupt. It probably isn't the most hospitable world for life in the Solar System, but it would probably be a geologist's or volcanologist's dream to poke around there.

Io is one of Jupiter's largest moons and potentially the most interesting. It is best known for its active volcanoes and a "rotten egg" smell that comes from the sulfur and sulfur dioxide being spewed from them.

TITAN

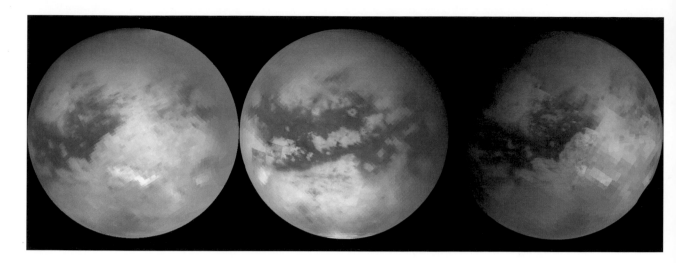

Until very recently, Saturn's moon Titan gave us the view of a hazy orange sphere that wasn't very exciting. NASA's Cassini mission changed all of that. In 2005, it launched a probe to Titan's surface to glimpse what lay below. As it plunged, the Huygens probe (named after a famous seventeenth-century Dutch astronomer) relayed data back to Earth about the atmosphere and weather it encountered. It revealed a world that looked much like our home planet, with surface features such as riverbeds, vast deserts covered in dunes, and even lakes. These lakes were filled not with water, however, but liquid hydrocarbons, much like the fuel we put into our cars. While this stuff would be deadly to us, it may not be to other forms of life that may have evolved there. The discovery of these toxic (for humans) pools was exciting because it was the first time open bodies of liquid had been found anywhere besides Earth.

Titan is the largest of the fifty-three known moons orbiting Saturn. Despite its distance from the Sun, Titan is arguably one of the most Earthlike worlds we have found to date. With its thick atmosphere and complex, carbon-rich chemistry, Titan resembles a frozen version of Earth several billion years ago, before life began pumping oxygen into our atmosphere.

POINTS TO PONDER

THE EARTH BELONGS TO AN INTERESTING COSMIC NEIGHBORHOOD CALLED THE SOLAR SYSTEM.

SOME OF THE SOLAR SYSTEM'S LESSER-KNOWN BUT STILL FASCINATING ATTRACTIONS CAN BE FOUND IN THE MOONS OF JUPITER AND SATURN.

WE CAN STILL ADMIRE PLUTO, EVEN IF IT'S NO LONGER CONSIDERED A "REAL" PLANET.

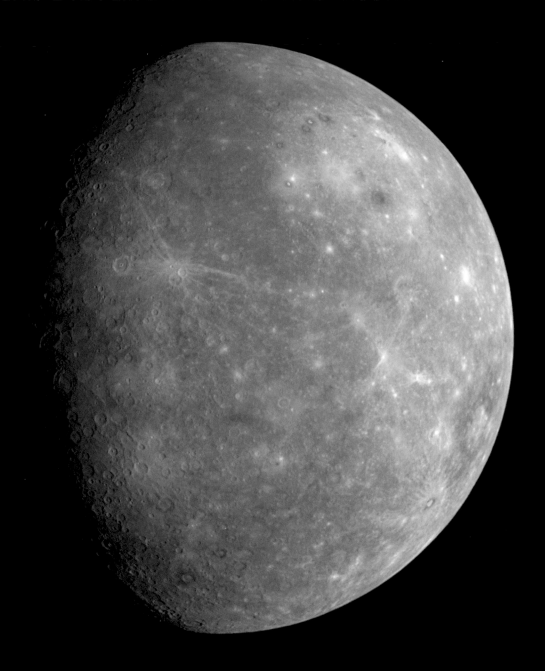

MERCURY: 3.22 LIGHT-MINUTES

Mercury is our smallest planet, with a diameter that spans only 3,000 miles (4,800 kilometers)—just a little more than the distance between New York and Los Angeles. It's a rocky world covered with pockmarks left over from countless impacts from meteors billions of years ago when the Solar System was still forming. Mercury moves around the Sun faster than any other planet, with its year lasting about eighty-eight Earth days. From what we know so far, Mercury might be more like our Moon than any other planet. Both are covered by a thin layer containing similar minerals, and they both have broad, flat plains, steep cliffs, and many deep craters.

VENUS: 6.01 LIGHT-MINUTES

Radar data collected over many years were used to create this beautiful, color-coded portrait of Venus that shows different levels of elevation on the planet. Venus has roughly the same size, mass, density, and composition as Earth. Until the 1960s, scientists speculated that Venus might have been very Earthlike, at one time home to lush tropical forests. That view changed when new observations confirmed a superheated surface with temperatures and pressures nearly a hundred times that of Earth. But the biggest difference with Earth lies within the atmosphere of Venus. Clouds on Venus are made not of water, as they are on Earth, but rather of concentrated sulfuric acid—essentially battery acid.

MARS: 12.7 LIGHT-MINUTES

Like Earth and the rest of our Solar System, Mars (the fourth planet from the Sun) is about 4.6 billion years old. Viewed from Earth, Mars is a bright reddish-orange. This color is similar to the color of rust. That's no accident: The Martian surface is covered with iron oxides, a more technical name for rust. The high, wispy clouds are mainly comprised of water ice. At the far left, clouds are seen around the peak of Olympus Mons, the largest volcano in the Solar System. At top center, the ice cap covering the Martian North Pole is visible. Scientists think that Mars was volcanically active for its first two to three billion years.

CURIOUSER AND CURIOUSER ABOUT MARS: 12.7 LIGHT-MINUTES

NASA's most recent rover on the red planet has sent a beautiful postcard from Mars. Taken just a couple weeks after the Curiosity rover landed on Mars on August 6, 2012, the foreground shows the gravelly area near Curiosity's landing site. In the distance is the base of the 3.4 mile-(5.5 kilometer) tall mountain called Mount Sharp, where the rover should eventually end up. The color scale has been enhanced in this photo to show the Martian scene in lighting conditions similar to what we have on Earth. This color enhancement helps the scientists who are actively analyzing the terrain from afar.

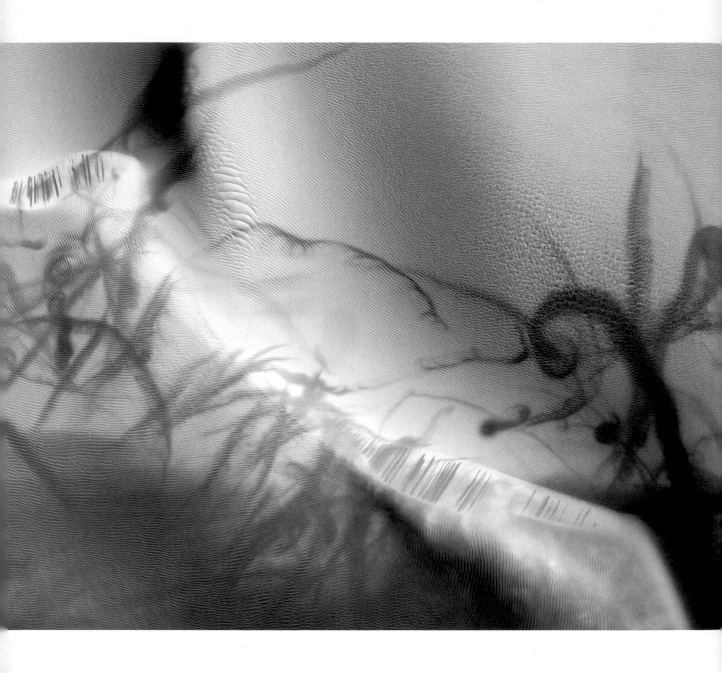

MARS DUST DEVILS: 12.7 LIGHT-MINUTES

While this image may look like someone's intricate tattoo, the dark swirls are, in fact, spectacular patterns made by winds in sand dunes on Mars. As strong winds blow across the Martian landscape, they generate and propel dust devils, which produce temporary dark scars across the lighter sand dunes. This image, taken with the Mars Reconnaissance Orbiter, shows an area of roughly about half a square mile (1.3 square kilometer) of the Martian surface.

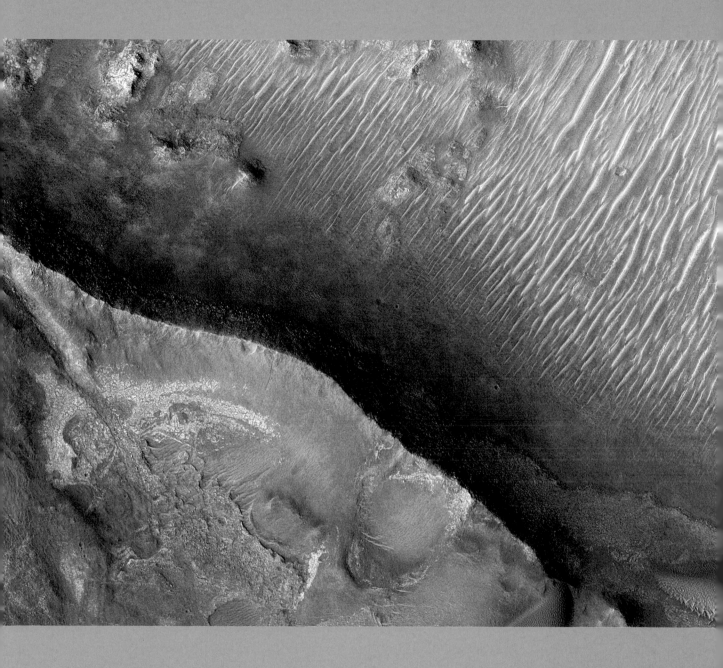

MARS NILI FOSSAE: 12.7 LIGHT-MINUTES

This impressive if strange-looking image shows a region of Mars that lies between a large volcano and an ancient impact basin. It is home to a collection of curved troughs cutting about 1,600 feet (500 meters) into the crust, one of them seen in this image. The European spacecraft Mars Express detected clay minerals there. That's good news for astrobiologists, since the clays point unmistakably to the presence of water at some point in the past. Even better, the minerals suggest that the water collected in pools on the surface, a seemingly comfortable habitat for life.

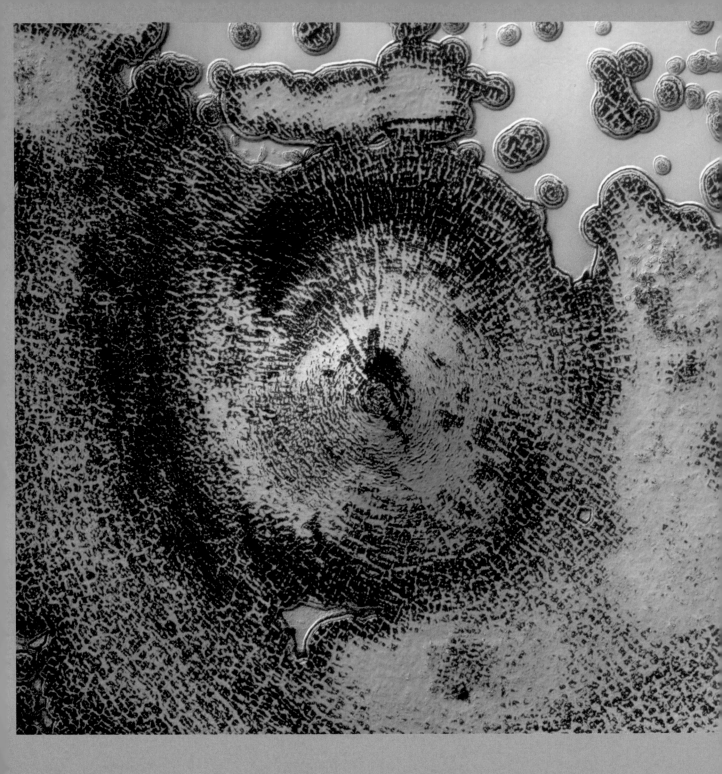

MARS SOUTH POLAR CAP: 12.7 LIGHT-MINUTES

Like Earth, Mars has frozen polar caps. But unlike Earth, the Martian polar caps are made from carbon dioxide ice as well as water ice. During summer in the Martian southern hemisphere, much of the carbon dioxide ice turns to vapor. Climate data from the Mars Express spacecraft show that the southern polar cap is actually built up every year by two different weather systems—one that produces carbon dioxide snow on the western side of the pole, and another in which only ground frost occurs on the eastern side. This means that the southern polar cap is asymmetrical during the summer when the ground frost turns to vapor more easily, leaving the snow-built cap to the west of the pole as all that's left.

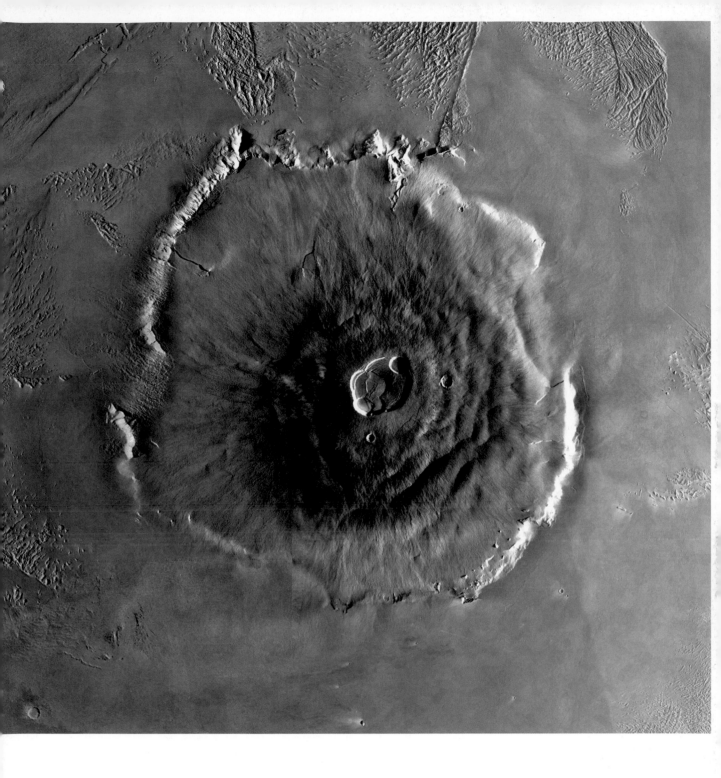

OLYMPUS MONS: 12.7 LIGHT-MINUTES

It's hard to tell from this top-down view from a spacecraft around Mars, but this image shows the tallest volcano—or mountain, for that matter—in the entire Solar System. This is Olympus Mons, a dormant volcano that rises a staggering 78,000 feet, nearly seventeen miles (about 24,000 meters) above the planet's surface. To give some perspective, that is more than 2.5 times taller than Mt. Everest. Why is it so big? Scientists think that it is supersized because Mars's lower gravity doesn't pull as hard on the growing volcano. Unlike Earth, too, where the surface plates often shift (as in an earthquake), the crust of Mars is stationary. This allows lava to continuously pile onto itself in one location, making one monster Martian volcano.

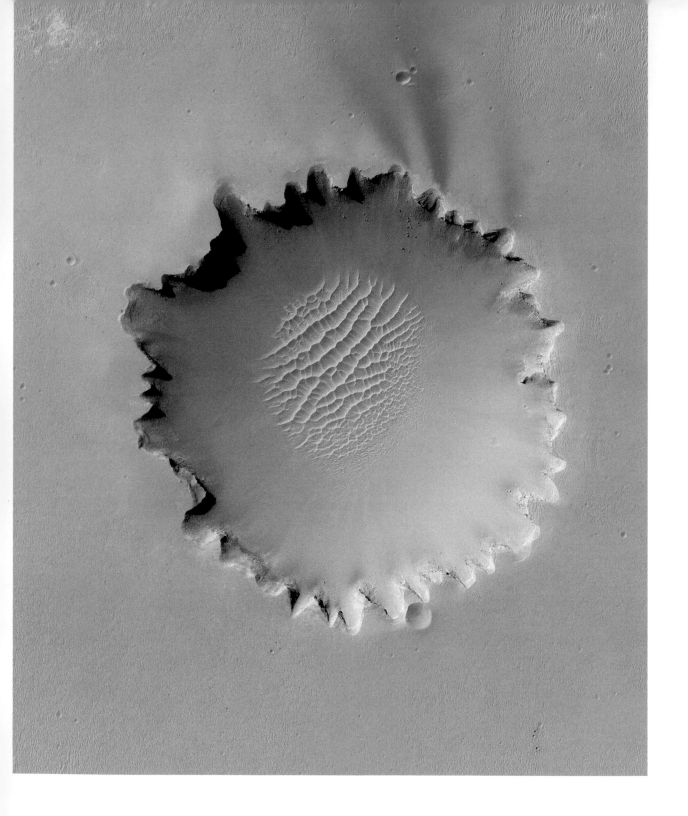

MARS VICTORIA CRATER: 12.7 LIGHT-MINUTES

It took nearly two years for NASA's rover Opportunity to get to Victoria Crater, seen here from above in an image from the Mars Reconnaissance Orbiter. The arduous journey was worth it, since Victoria Crater was the largest crater—roughly the size of a football stadium—explored by either Opportunity or its sister rover, Spirit, to date. Opportunity spent about a year exploring and studying this impact crater, which is near the Martian equator. The rover can actually be seen in this image as a small dot in the upper left (or 10 o'clock position), along the crater rim.

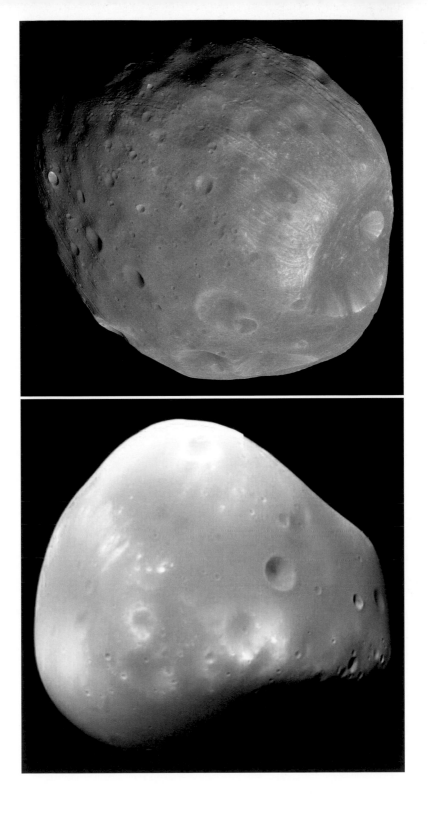

PHOBOS AND DEIMOS: 12.7 LIGHT-MINUTES

Mars has two relatively small moons known as Phobos (top), named after the Greek god whose name means "fear," and Deimos (bottom), after the Greek god of terror. Unlike other moons around the Solar System—including our own—this pair has one obvious difference: They are not round. Their potato-like shapes and differing speeds and orbits lead scientists to believe that they are actually captured asteroids.

JUPITER: 43.3 LIGHT-MINUTES

Jupiter is the largest planet in the Solar System and the fifth planet from the Sun. Its diameter is more than eleven times that of Earth, or about one-tenth that of the Sun. This psychedelic image of Jupiter has been color coded to show cloud height from high altitude (white) through midrange altitude (blue) to low altitude (red). The Great Red Spot and its neighbor, "Red Spot Junior," are at the top of the atmosphere, explaining why they appear mainly as white in this image.

JUPITER'S GREAT RED SPOT: 43.3 LIGHT-MINUTES

The Great Red Spot, a vast storm in Jupiter's southern hemisphere, rotates counterclockwise, in the opposite direction from the clockwise one that storms in the Southern Hemisphere follow on Earth. The colors in the Great Red Spot change as different chemicals are churned from the bottom layers up to become the top ones. Winds at the edge of the spot tear along at up to 350 miles per hour (560 kilometers per hour). The Great Red Spot is about three times the size of Earth, so big that you can see it through a small backyard telescope if the conditions are right. People have been watching this giant Jovian storm through telescopes big and small now for hundreds of years.

IO, EUROPA, GANYMEDE, AND CALLISTO: 43.3 LIGHT-MINUTES

Jupiter's four biggest and potentially most interesting moons are named (clockwise from top left) Io, Europa, Ganymede, and Callisto after figures in classical Greek mythology. Io is covered in sulfur-spewing volcanoes. Europa's surface is mostly water ice and may harbor a liquid ocean underneath containing two to three times as much water as Earth does. Ganymede is the largest moon in the Solar System (bigger than the planet Mercury), and it is the only moon known to generate its own internal magnetic field. Two sulfurous eruptions are visible on Jupiter's volcanic moon Io (the small purple plume on the left and the small purple swirl at top). Callisto's surface is extremely heavily cratered and ancient, which provides a record of events from the early history of the Solar System.

SATURN: 1.32 LIGHT-HOURS

Saturn is the second largest planet in the Solar System, surpassed in size only by Jupiter. Saturn's famous ring system glows with scattered sunlight in this image, made by the Cassini spacecraft as it passed behind the planet in 2006. Saturn has almost fifty diverse moons that range in size from just under two miles wide to about the width of the continental United States. This image also contains our home planet, the white dot between the bright main rings and the thinner gray-brown ring in the upper left. The image looks different from more typical views of Saturn because it was made by combining layers of infrared, visible, and ultraviolet light.

ENCELADUS: 1.32 LIGHT-HOURS

Enceladus is a small moon of Saturn, only about as wide across as the United Kingdom, with a surface covered with intriguing scars. What are all these marks? Scientists think they are a giant system of fractures in the ice. The fractures are coated with organic materials—in other words, the stuff necessary to make life as we know it on Earth. Interestingly, the south polar-region, where many of the fractures are found, is the hottest place on Enceladus. That's like finding out that the Antarctic region on Earth is actually hotter than our tropics.

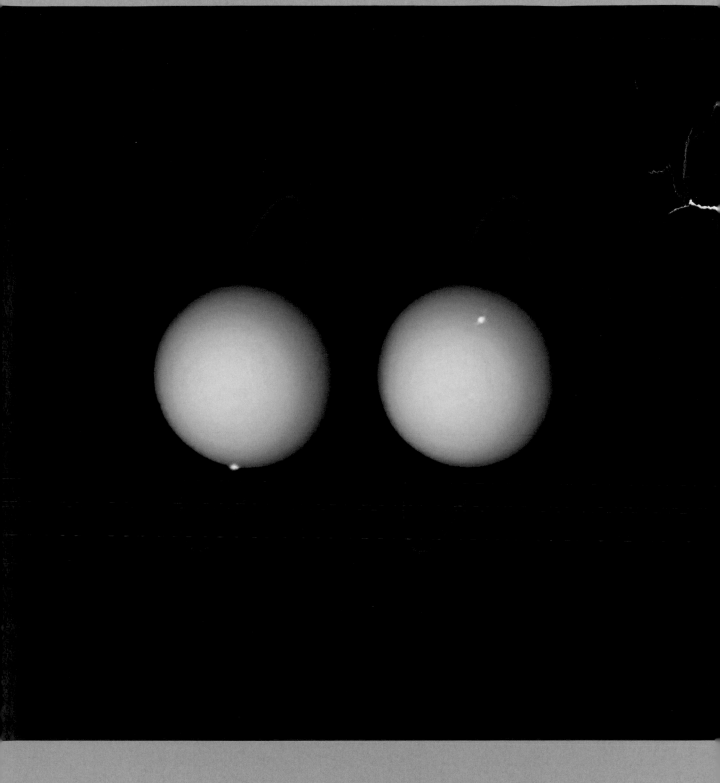

AURORAS ON URANUS: 2.7 LIGHT-HOURS

Some of us have been lucky enough to see the Aurora Borealis, or Northern Lights, or the equivalent sky show from south of the equator. But other planets have auroras, too. Scientists have seen auroras on Jupiter, Saturn, and, as we can see in these images, Uranus. Taken with the Hubble Space Telescope in 2011, these images show the auroras of Uranus as they appear in visible and ultraviolet light. The data of the auroras from Hubble have been combined with photos of the whole planet taken by the Voyager 2 spacecraft during its flyby of the planet in 1986, along with optical data from the Gemini Observatory in Hawaii, to give us this unique shot of this enigmatic planet.

COMET HALE-BOPP: 656 LIGHT-SECONDS

Why are there two tails behind this comet? When a comet comes close to the Sun, it begins to heat up and lose dust and gas. The dust that evaporates from the comet follows the comet's orbit while reflecting the Sun's light, thus appearing white. The blue tail shines due to gas from the comet, primarily carbon monoxide, interacting with particles emitted by the Sun.

COMET C/2001 Q4 (NEAT): 160 LIGHT-SECONDS

Because of their shape and their path through the Solar System, many people probably think that comets are akin to "shooting stars" or other projectiles seen due to friction in our atmosphere. However, there is no atmosphere in space, and comets are not actually burning up. Instead, comets are visible on Earth because they scatter sunlight. As they are warmed by the Sun on their path inward from the far reaches of the Solar System, these icy balls of dust and rocks begin to shed their material.

5

THE BIRTH AND LIFE OF STARS

A host of other stars...so numerous as almost to surpass belief.

— GALILEO GALILEI

WITH OUR SPIN AROUND THE SOLAR SYSTEM COMPLETE, let's keep going to our next destination: the stars. If you live in an urban or suburban environment, you can probably see a few dozen or even a few hundred stars on a cloudless and moonless night. If you can get to a spot that is far away from city or other artificial lights, you may expect to see up to a few thousand stars.

While a dense, starry sky is a spectacular sight to behold, as Vincent Van Gogh captured in his painting *Starry Night*, that impressive canopy represents just a drop in the bucket in terms of the number of stars out there. Stars are the building blocks of galaxies, and our Milky Way galaxy contains an estimated 200 billion to 400 billion stars.

PREVIOUS SPREAD:
A starry evening in Cape Schanck, Victoria, Australia shows bands of dust and glowing nebulas, along with some of the billions of stars that make up our Milky Way galaxy.

Van Gogh's painting Starry Night.

That's a lot of stars. Their ubiquitous presence and the staggering amount of stars are good motivation to understand how they work. There are many other reasons, however, including having a better handle on how our own local star (the Sun) operates and may behave in the future.

This brings us to one of our first and most important points about stars: they are not around forever. Instead, they are akin to humans. That is, they are born, live their lives, and ultimately die.

Exactly how the stars live their lives depends on certain important characteristics. As with people, there can be a lot of complexity, determined by many factors, including what they are made of and the environment they live in.

Let's start with the basics.

Stars come in different colors, which often dictate how hot they are. Contrary to our experiences with kitchen stoves, "red" does not always mean the highest temperature. In fact, blue and white stars are the hottest, while yellow stars are medium hot, and orange and red stars are generally cooler. For reference, our Sun is a yellow star, which makes it average on the stellar temperature scale.

The Sun is the nearest star. By studying the Sun, which is close enough that we can see things in great detail, we learn important information about other stars that are much farther away from us.

Another critical trait of any star is its mass, or size. As a rule of thumb, the larger the star, the shorter its life will be. Once again, our Sun falls in the middle of the pack and is a medium-sized star. This means it will probably live for about 10 billion years. Since astronomers estimate that the Sun has been around for about 5 billion years, this also means that it is literally middle-aged. In contrast, the smaller, more fuel-efficient stars can live several times longer than the Sun, while the biggest stars will burn through their fuel supply in only a few million years.

Astronomers have come up with a sophisticated way to classify stars by their color, luminosity, size, stage of life, and other characteristics. The Sun, for example, is classified as G2V, the G meaning that it is a yellow star, the 2 indicating its surface temperature (about 9900 degrees Fahrenheit, or 5500 degrees Celsius), and the V indicating that the Sun is a main-sequence star that generates its energy by a nuclear reaction. Unless you plan on studying stellar evolution in great detail, however, we suggest you don't worry too much about these classifications.

THE CONSTELLATIONS

For many thousands of years, humans have gazed up at the heavens and discerned patterns in the sky of the white dots of light. We ourselves have spent more than a few nights looking at the sky and trying to find those familiar patterns. These patterns, called constellations, have been used since early civilizations to explain cultural stories, offer explanations for such things as the changes in seasons, and serve as special markers to help navigate ships on the seas.

With the help of modern technology, we now know that the stars that make up a constellation are not necessarily near one another out in space. Sometimes they just appear to be next to

The constellation of Ursa Major is a well-known sight in the Northern Hemisphere. Although it is often called the Plough or the Big Dipper, most mythologies have thought of it as a great bear, with the three stars on the left being its tail and the four at the right part of its back. (Ursa Major means "great bear" in Latin.) The two stars on the far right, called "the pointers," trace a line that can be used to find Polaris (the north star), a trick travelers have been using for thousands of years.

each other from our vantage point on Earth, but they can really be billions of miles apart in their distances from us.

Today, we have eighty-eight "official" constellations as defined by the International Astronomical Union. Most of the modern constellations recognized come from early Greek, Roman, and Egyptian mythology. But nearly every ancient culture—from Arabia to China to Native Americans—has its own names for the constellations and stories to go with them.

Did you know that the constellations of the zodiac shift over time? That's because the Earth wobbles ever so slightly on its axis and over the course of a few thousand years, the stars overhead are not exactly where they used to be. So though you might be a Scorpio today, sometime down the road, you might end up a Sagittarius.

STELLAR BIRTH

Where do stars come from? The basic story remains the same for most stars, including the one we told about the Sun in the previous chapter. Stars are born when giant clouds of gas and dust collapse. As the cloud condenses, a core of hot material begins to form at the center. This is the baby star, and the remaining material that isn't swept into this core can become planets, asteroids, and other debris. This might sound as if it's a simple story, but there are actually many details about the birth of stars that astronomers have been and are still trying to figure out.

Once this multimillion-year birthing process is over, the star burns, or shines, through a process known as nuclear fusion. An atomic core is called a nucleus. Nuclear fusion occurs when two atomic cores join together, or fuse. When this happens, the two cores become a single, heavier core. At the same time, a large amount of energy is released.

It is this process of atomic nuclei smashing together and releasing energy that powers a star. This is what we mean when we say that a star is "burning." For most of a star's life, this burning involves hydrogen—the lightest and most abundant element in the Universe—being fused together to make helium. This is the stage that our Sun is in now.

ABOVE:

This artist's illustration shows what a massive baby star with a dusty disk closely encircling it might look like.

RIGHT:

As a star nears the end of its life, heavy elements are produced by nuclear fusion inside of it, such as iron, labeled in the center of this schematic. These elements are concentrated toward the star's center.

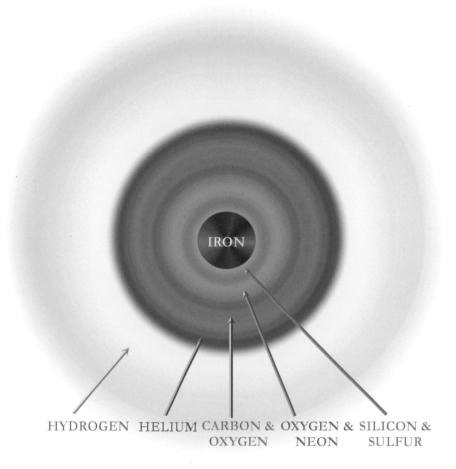

IRON

HYDROGEN   HELIUM   CARBON &   OXYGEN &   SILICON &
                             OXYGEN   NEON   SULFUR

Stars in space are often not found alone. Rather, they travel with their own stellar entourages in what are known as star clusters.

There are two basics types of star clusters: globular and open clusters. Globular clusters are found on the outskirts of galaxies and have a tight packing of thousands or even millions of mainly very old stars. Open clusters, on the other hand, are not as densely crowded, and they contain more young stars.

STARS HANG OUT WITH OTHER STARS

In the center of 30 Doradus, a region where many stars are being born, lies a big cluster of large, hot, and massive stars. These stars, known collectively as star cluster R136, were imaged in optical light by the Hubble Space Telescope.

OLD STARS

Things get a little more interesting once a star has no more hydrogen left to fuse in its core (this won't happen to our Sun for about 5 billion more years). With the hydrogen in the core gone, the energy stops being produced there. This causes the core to slowly collapse and heat up.

The nuclear fusion continues, however, by moving to where there is still hydrogen. This means that nuclear fusion starts up in a hydrogen-filled shell of gas that is outside the core, and this then powers the star. This is when a star enters what astronomers call a "red giant" phase.

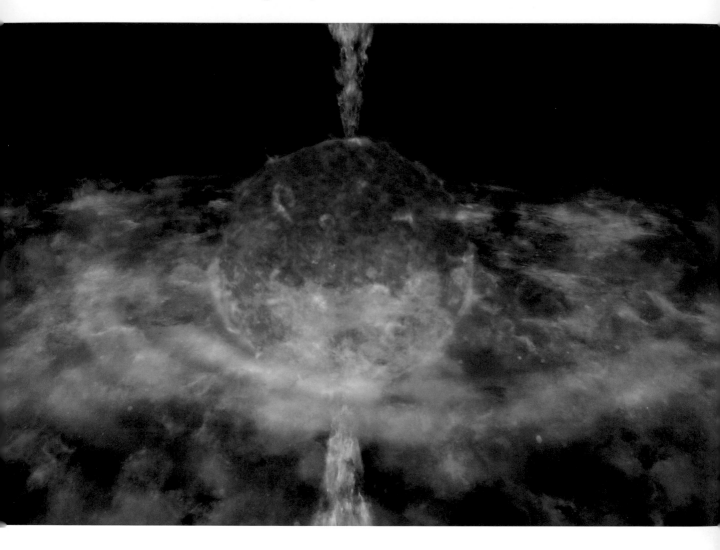

This artist's illustration shows a close-up of how a red giant star would look surrounded by a dusty and gaseous disk with a pair of jets blasting out of the system.

After the hydrogen is used up in this shell, the nuclear fusion process in the red giant star begins to join the newly created heavier atoms of helium together. These fused helium atoms become even heavier elements as the nuclear fusion process moves down the periodic table of elements until it gets to carbon.

If a star isn't big enough to generate a temperature high enough to fuse carbon, the star gets a little stuck. Instead of combining atoms together to make new ones, the carbon and oxygen begin to build up in the core. As the star's core attracts more and more mass, it heats up. This causes the outer layers of the star to puff out.

This red-giant phase doesn't always happen in a smooth and consistent process. Instead, as the nuclear fusion in the core starts to sputter like a car running out of gas, the star sheds material in uneven fits and spurts.

Astronomers have captured many images of stars in this phase, which they call "planetary nebula." This can be utterly confusing because planets are nowhere to be found in this discussion. Planetary nebulas have this name only because of a historical relic, that is, they resemble a planet with a disk around it as seen through a low-powered telescope, which is all astronomers had access to until the last few decades.

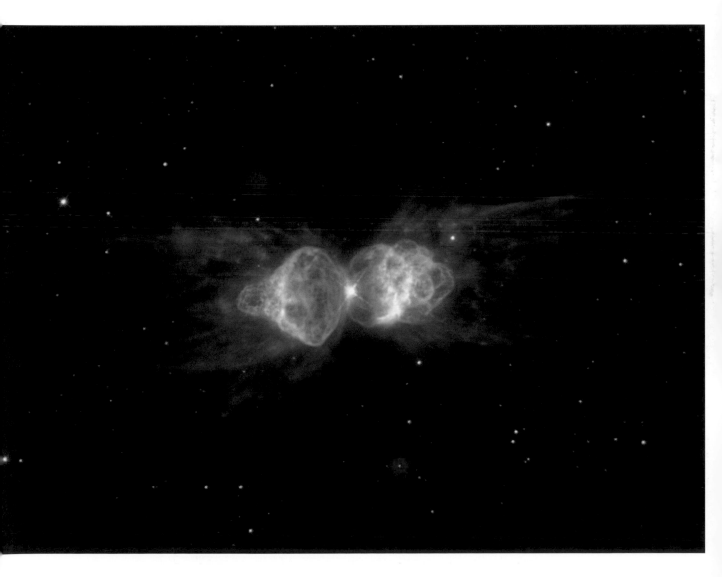

The Ant Nebula exhibits a typical bipolar shape of a planetary nebula when viewed from the side. A Sun-like star has shed material from its outer layers until its core is exposed, releasing light that makes the gas glow. This image combines optical light from the Hubble Space Telescope (green and red) and X-ray light from the Chandra X-ray Observatory (blue).

Today, there is no confusion among scientists that these objects represent the dying throes of a star and have nothing to do with planets. When our Sun reaches this point—again, billion years of years from now, so not to worry—it will swell. Not just a little, but quite a lot. In fact, astronomers estimate that when this happens, the Sun will puff out so far that it will engulf the inner planets of the Solar System: Mercury, Venus, and maybe even Earth. Our Sun, and other stars of its kind, will continue to exist as a red giant until there are simply no more outer layers to release. Left behind will be a very dense, small core, or what astronomers call a white dwarf. With no fuel to burn, the white dwarf will just sit in space, radiating heat for billions of years. At some point, the star will eventually go from a glowing cosmic ember to a dark one.

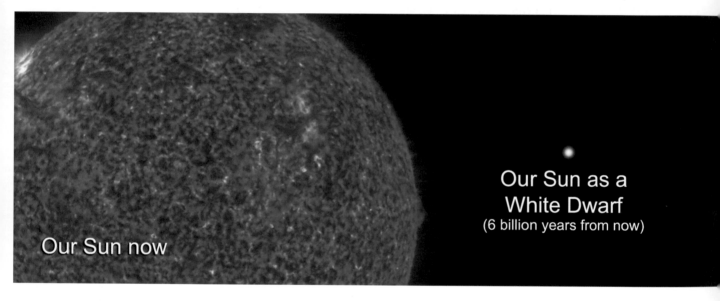

Our Sun as a
White Dwarf
(6 billion years from now)

Our Sun now

This illustration shows a relative comparison of the Sun as it is now, and what it might look like in, say, six billion years as a white dwarf.

The slow march toward becoming a white dwarf is the gradual and unassuming fate for stars like our Sun or smaller, but what about the bigger ones? The short answer is that they don't go quietly. These stellar big boys and girls could be considered the true party animals of the cosmic scene. That's because their nuclear fusion doesn't stop with nuclear fusion at carbon. They keep going, making heavier and heavier elements until they get to iron.

When the star starts fusing iron, it's in serious trouble. That's because instead of releasing energy like other atoms do when they join, two iron atoms that are being fused together actually suck up energy. This creates a runaway chain reaction that makes the star cool and the pressure precipitously drop. This ultimately leads to a dramatic collapse of the entire star.

After the star collapses onto the core, it rebounds and the outer layers of the star go hurtling into space. Astronomers call these explosions supernovas. When they go off, supernova explosions release so much energy that they can outshine an entire galaxy. That's no mean feat when you take into account that there are billions of stars in a galaxy.

When these massive stars go through a supernova explosion, they still leave behind a core of what was once the star. This core, however, is now even denser, since the collapse of the core has continued until all of the electrons and neutrons are packed together like a crush of people trying to get through an exit with nowhere to go. This squeezed-down object is what scientists call a neutron star, and it is so dense that just one teaspoon of it weighs more than a billion tons. Put another way, a neutron star packs the mass of something the size of our Sun into a sphere about as wide as Manhattan.

All supernovas are the result of stars that have exploded. But there are different ways for a star to explode, and astronomers are very diligent in trying to classify how each star goes out. Some stars become supernovas because they are simply too massive and collapse down on themselves when they run out of fuel.

This illustration shows just one possible path a star might take over the course of its lifetime, changing from a Sun-like star (bottom left) to red giant (top middle) to planetary nebula (far right), and eventually, ending as a white dwarf.

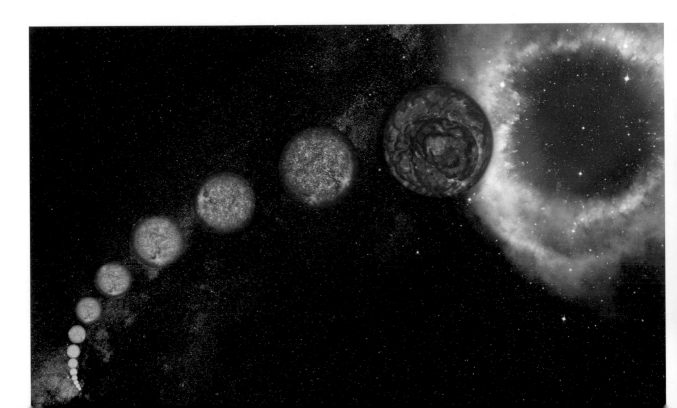

Another kind of supernova happens when a white dwarf has a companion star in close orbit around it. If the white dwarf is big enough, it will pull material from the companion star using its gravitational prowess to its own surface. Eventually if the white dwarf collects too much stolen material onto its surface, it will trigger a thermonuclear explosion and the whole star is blown apart. Astronomers find this type of supernova very handy because they think they explode with the same output of light every time and can be used for measuring vast distances across the Universe. (In Chapter 8, we will discuss how they are being used to study the mystery of dark energy.)

WE ARE THE STUFF OF STARS

What relevance do the life and death of stars have to life here on Earth? The answer is: Everything.

All of the elements we need to live—the oxygen we breathe, the calcium in our bones, the iron in our blood—were forged in the furnaces of previous generations of long-lost stars. This is not an exaggeration. All of these life-giving building blocks are here on Earth because they were swept up by our Sun's prenatal cloud and incorporated into our planet as it developed. Every time a star releases its fused elements into space, either through the gentle breeze of a red giant or the violent blast of a supernova, it is actually enriching the next generation of stars and planets with these essential elements.

Supernova remnants—that is, the debris field left behind after the explosion—are like snowflakes: No two are ever exactly the same. To our eyes, they are also some of the most spectacular images in astronomy. But what exactly are we seeing?

Because many of these images are composites—that is, they have several different layers of light stacked together—it can be useful to break them down into their component layers to get a better look. One famous supernova remnant is named after Johannes Kepler, a famous German astronomer who lived about four hundred years ago. When Kepler (and others) saw the star explode back at the beginning of the seventeenth century, he saw a bright light. Today, however, we can see much more—especially when we look at the aftermath of the supernova explosion in X-ray light.

By assigning red to the lower range of X-rays, green to the medium range, and blue to the most energetic X-rays, we can layer the slices together to get one three-color image. This type of layering allows our humans eyes to see what they cannot on their own.

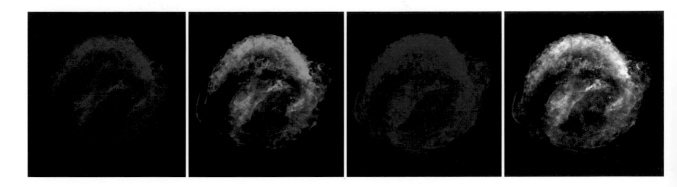

Another example of layers involves different types of light coming from the Crab Nebula. The Crab is one of the most famous objects in space. It first showed up in the night sky almost a thousand years ago, and practically every space telescope ever built has looked at the Crab. Here, we'll show you what happens when you create a composite image by combining optical, infrared, and X-ray light. While we like the Crab in any kind of light, we think this composite version finally begins to tell us the whole story at the far right.

ABOVE:

Kepler's supernova remnant in X-ray light. Three slices of X-rays (low, medium and high energies) are added up to create the X-ray composite image on the far right.

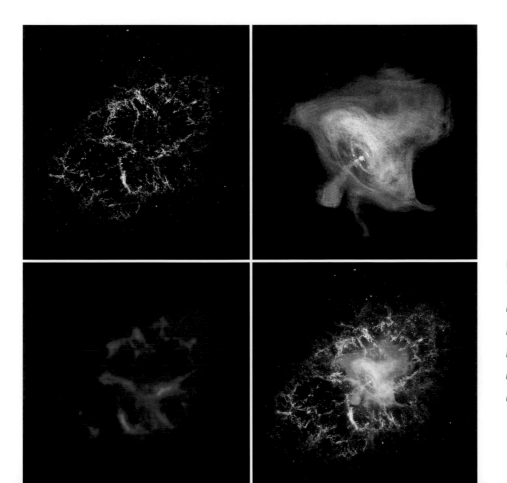

LEFT:

The Crab Nebula in optical, X-ray, and infrared light. The three kinds of light were added up to create a composite image on the lower right.

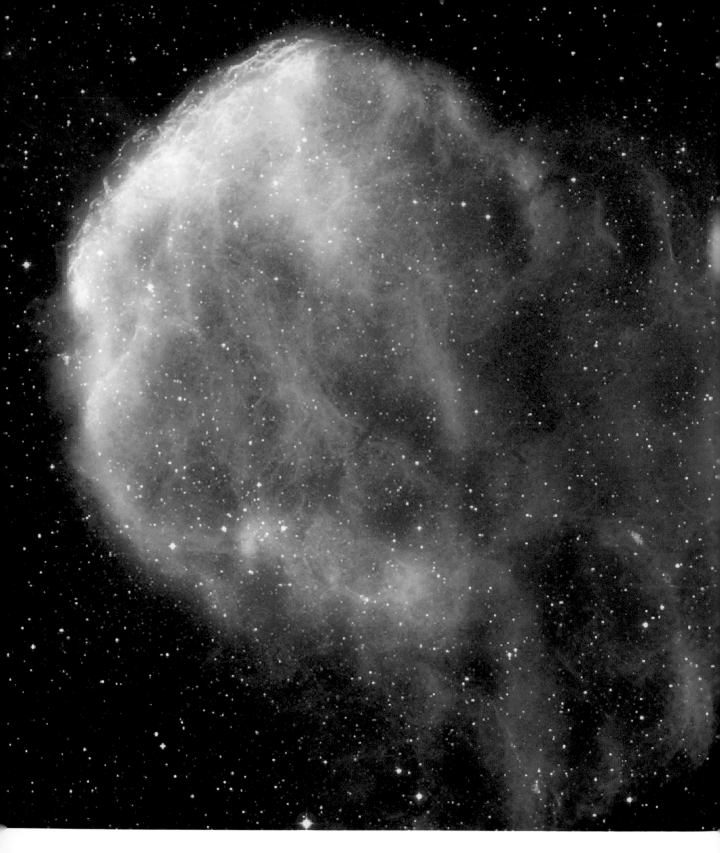

IC 443 is a supernova remnant with a neutron star embedded inside it. Three different kinds of light are in this composite, including X-ray (blue), radio (green), and optical (red). The location and orientation of the neutron star's wake (the brighter blue spot in the lower middle part of the image) are mysterious for astronomers who would have expected it to be aligned toward the center of the supernova remnant.

BLACK HOLES:
SMALL, MEDIUM, AND LARGE

When a star is extremely gigantic, its fate is even more bizarre than going out in a blaze of supernova glory and creating a neutron star. The very largest of stars will just keep on collapsing until it forms a black hole. Once a star collapses into a black hole, nothing—not even light—can escape from its boundaries. We know about these extreme objects because we see material around the black hole and the effects the black hole has on its environment through its intense gravity.

Scientists have speculated for hundreds of years that some stars might collapse and become dark, that their gravity might be so strong that nothing, not even light, can escape. These dark stars are called black holes, as illustrated here. Since no light of any kind can escape a black hole, we cannot directly image them. But one way to find them is to search for high-energy radiation from a disk of hot gas swirling toward a black hole using telescopes that can detect X-rays.

Black holes have been the stuff of science fiction for a long time, but astronomers have hard evidence that they do actually exist. Actually, we know about lots of black holes. In fact, astronomers think they come in two, if not three, distinct sizes. When giant stars run out of fuel and collapse on themselves, as we have just discussed, they form the smallest kind of black hole known, called a "stellar-mass black hole" or "stellar black hole."

These black holes are probably pretty common—or at least as common as the very biggest stars are. Astronomers have used telescopes to carefully study some of the nearest ones and we now have a good handle on some of their characteristics like how much they weigh and how fast they are spinning.

Likewise, astronomers have suspected for a while that most galaxies—including our own Milky Way—contain enormous black holes at their centers. We call these beasts "supermassive black holes." We will discuss these galactic monsters in more detail in the next chapter.

Scientists have recently found evidence that there is a medium-sized class of black holes. While the data on these middleweights are not as ironclad as they are for the other two categories of black holes, many people are on the lookout for this midsize kind of black hole. Astronomical theorists are also debating how these medium-sized black holes form. While the jury is still out, odds are that the death of stars is somehow involved in this ongoing mystery—which is yet another reason to be fascinated with stars.

LOOKING THROUGH TIME

This is a good place to stop for a moment and discuss one of the most confusing, yet intriguing, aspects of exploring space. All of the space images—including stars—that you see are snapshots in time, which means that you are looking back in time.

Let's compare this with a more familiar scenario first: in photos on Earth, the time that has passed from someone's baby pictures to adulthood is just a couple of decades. If you were not in contact with that person between their infancy and their adulthood, the only view you would have of him or her would be that baby picture taken decades before.

Images from space are just like baby pictures except that these cosmic snapshots represent a much longer period of time, usually millions—if not billions—of years. The light from that galaxy has taken that many years to travel to reach our eyes or telescopes. That's how we see "back in time" with these images, just as we see into the past, traveling through time, when we look at old family photos.

But if these astronomical images are snapshots from the past, then how do we know what's happening with that object today? The answer is: We don't. Just as you wouldn't know what happened as an adult if you didn't keep in touch with your friend after his or her toddler years by looking at baby pictures, we don't know what's happening with these objects at this very moment at that far-flung point in space.

Think of it this way: The Sun is less than ten light-minutes away, and it takes light from Jupiter a little more than half an hour to reach us. The nearest star outside our Solar System, as we mentioned is about four light years away and so the "latest" image of Proxima Centauri always shows us what happened four years ago. Earth is 26,000 light-years away from the center of our home galaxy, the Milky Way, so any information on stars or other objects there comes from light that actually left just about the time the Neanderthals went missing from the fossil record.

LIGHT ON THE STARS

STARS HAVE FINITE LIVES. THEY ARE BORN, LIVE FOR A PERIOD OF TIME, AND ULTIMATELY DIE.

IN GENERAL, THE SMALLER THE STAR, THE LONGER IT WILL LIVE. THE MORE MASSIVE A STAR IS, THE SHORTER ITS LIFE AND THE MORE SPECTACULAR ITS DEATH.

ALL OF THE ELEMENTS WE NEED TO LIVE—THE OXYGEN WE BREATHE, THE CALCIUM IN OUR BONES, THE IRON IN OUR BLOOD—WERE FORGED IN THE FURNACES OF PREVIOUS GENERATIONS OF LONG-LOST STARS.

ANTARES: 600 LIGHT-YEARS

This colorful vista is the region around the red supergiant star called Antares, seen as the bright yellow star to the lower left. This huge star is about seven hundred times the diameter of our Sun. If we had a star of this size in our Solar System, it would completely engulf all the planets as far out as Mars. The colorful backdrop behind the star Antares comes from pockets of gas (pink) and dust (yellow).

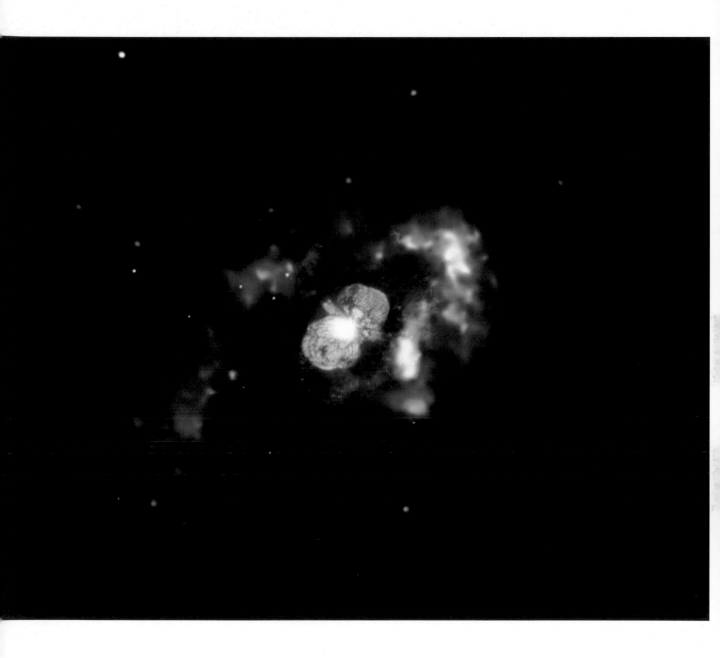

ETA CARINAE: 7,500 LIGHT-YEARS

A betting astronomer might put money down that Eta Carinae will be the next nearby star to explode. At a whopping 100 to 150 times the mass of the Sun, Eta Carinae is racing through its fuel. This image shows what Eta Carinae looks like in optical light (blue) and X-ray light (gold). Scientists think that when Eta Carinae does explode—either tomorrow or in the next several hundreds of years—it could become a really bright object in our night sky. Since it is about 7,500 light-years away from us, Earth will be safe from any harmful side effects of the explosion.

ORION NEBULA: 1,500 LIGHT-YEARS

Orion is one of the best-known constellations visible from the Northern hemisphere. One of the telltale ways to find Orion is his belt. The three stars from the lower left to the upper right in this image are named Alnitak, Alnilam, and Mintaka.

NGC 3603: 20,000 LIGHT-YEARS

The star-forming region NGC 3603 contains a massive young star cluster bathed in gas and dust within our Milky Way. Scientists think that the stars in this cluster formed during a big rush of star formation about a million years ago. The hot stars in the core of the star cluster are responsible for carving out a huge cavity in the gas.

BUTTERFLY NEBULA: 3,700 LIGHT-YEARS

The Butterfly Nebula (also known as IC 1318) gets its name from its two-winged appearance. It is a huge cloud of gas and dust with hot young stars forming within it. Hydrogen (green), oxygen (blue), and sulfur (red) depict a colorful and chaotic scene. A dark lane of dust cuts through the center.

ROSETTE NEBULA: 5,000 LIGHT-YEARS

The Rosette Nebula is a cosmic cloud of gas and dust that, as the name implies, resembles a flower. This view shows a long stem of glowing hydrogen gas. At the edge of a large gas cloud, the petals of this rose are actually a stellar nursery that gets its shape from winds and radiation from its central cluster of hot young stars.

EAGLE NEBULA: 7,000 LIGHT-YEARS

The Eagle Nebula is one of the most famous images in all of astronomy, thanks to an iconic image from the Hubble Space Telescope originally released in 1995. This version contains a wider field of view than the first Hubble image and was taken by the Kitt Peak National Observatory in Arizona. Right in the middle are dust columns that became known as the "Pillars of Creation." Here we see they are just part of a larger hollow shell of star formation, with a young star cluster at its center. The colors represent light given off by glowing hydrogen (green), oxygen (blue), and sulfur (red) detected in visible light.

REFLECTION NEBULA: 500 LIGHT-YEARS

A long tail of interstellar dust shines in the reflected light of nearby stars in this view of a nebula in the constellation Corona Australis (the southern crown). In some parts the dust accumulates to form dense clouds of gas from which it is thought young stars are born.

TARANTULA NEBULA: 160,000 LIGHT-YEARS

30 Doradus is a massive region of star birth. It is also called the Tarantula Nebula because its glowing filaments look like spider legs. Thousands of massive stars in the nebula are creating intense radiation and powerful winds. The Chandra X-ray Observatory detects gas that has been heated to millions of degrees by these stellar winds and by supernova explosions. These X-rays, colored blue in this composite image, come from shock fronts—similar to sonic booms—formed by this high-energy stellar activity. This hot gas carves out gigantic bubbles in the surrounding cooler gas and dust shown here in infrared emission from the Spitzer Space Telescope (orange).

CARINA NEBULA: 7,500 LIGHT-YEARS

The Carina Nebula is an immense landscape of dark dust columns silhouetted against glowing gas clouds found in the Milky Way. This nebula, almost 500 trillion kilometers wide, is both illuminated and sculpted by the intense radiation of its brilliant young stars.

E0102-72.3: 190,000 LIGHT-YEARS

The debris from the exploded star E01012-72.3 is located in the Small Magellanic Cloud, one of the nearest galaxies to the Milky Way. It was created when a star that was much more massive than the Sun exploded, an event that would have been visible from the Southern Hemisphere of the Earth a thousand years ago. Modern X-ray and infrared telescopes reveal an outer blast wave produced by the supernova (blue) and an inner ring of cooler (red-orange) material. A massive star (not visible in this image) is illuminating the green cloud of gas and dust to the lower right of the image.

HELIX NEBULA: 690 LIGHT-YEARS

The Helix Nebula is the closest and most impressive examples of a planetary nebula. From our vantage point on Earth, the Helix's discarded gases resemble a doughnut shape. Evidence, however, suggests that the Helix actually consists of two disks of gas nearly perpendicular to each other—picture one layer of gas coming out toward you. An unseen companion star may be the cause of this complex structure.

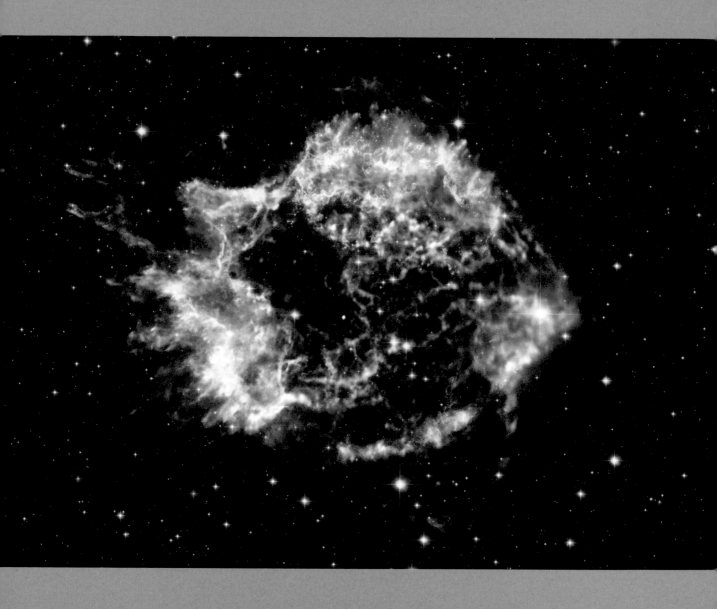

CASSIOPEIA A: 11,000 LIGHT-YEARS

More than three hundred years ago, the first light from an exploded star in the constellation Cassiopeia reached Earth. Dust and gas in our Milky Way galaxy obscured the explosion quite a bit, so reports of the event are sparse and hard to validate. In recent years, the remnant of the star has definitely been on astronomers' radar and today is called Cassiopeia A. This X-ray image shows the intricate structure that was created—like a psychedelic snowflake—when layers of different elements in the star were ejected and heated during the explosion.

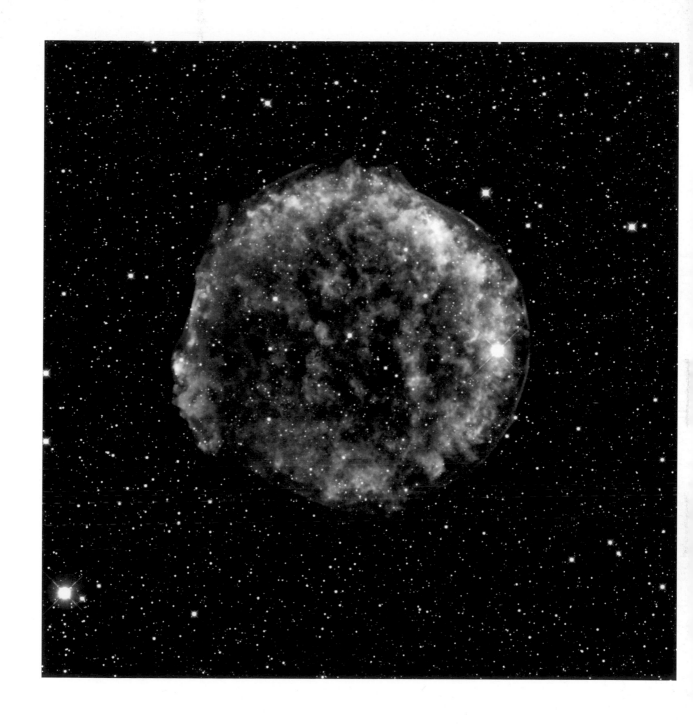

TYCHO'S SUPERNOVA REMNANT: 7,500 LIGHT-YEARS

In 1572, the Danish astronomer Tycho Brahe saw the explosion of a star that became known as Tycho's supernova. More than four centuries later, the explosion has left a blazing hot cloud of expanding debris (green and yellow). Material moving at six million miles per hour (about 10 million kilometers per hour) has created two shock waves that glve off X-rays, one moving outward into gas outside the star (seen as a blue sphere), and another moving back into the debris. These shock waves produce sudden large changes in pressure and temperature, like an extreme version of sonic booms produced by the supersonic motion of airplanes.

G292.0+1.8: 20,000 LIGHT-YEARS

This X-ray image of the supernova remnant G292.0+1.8 shows a rapidly expanding shell of gas containing important elements such as sulfur and silicon (blue), oxygen (yellow and white), and magnesium (green). These elements, which are critical to our existence here are Earth, were created both when the star was still living and in the explosion itself. Explosions like this one dispersed elements that were necessary to form our Sun and Solar System.

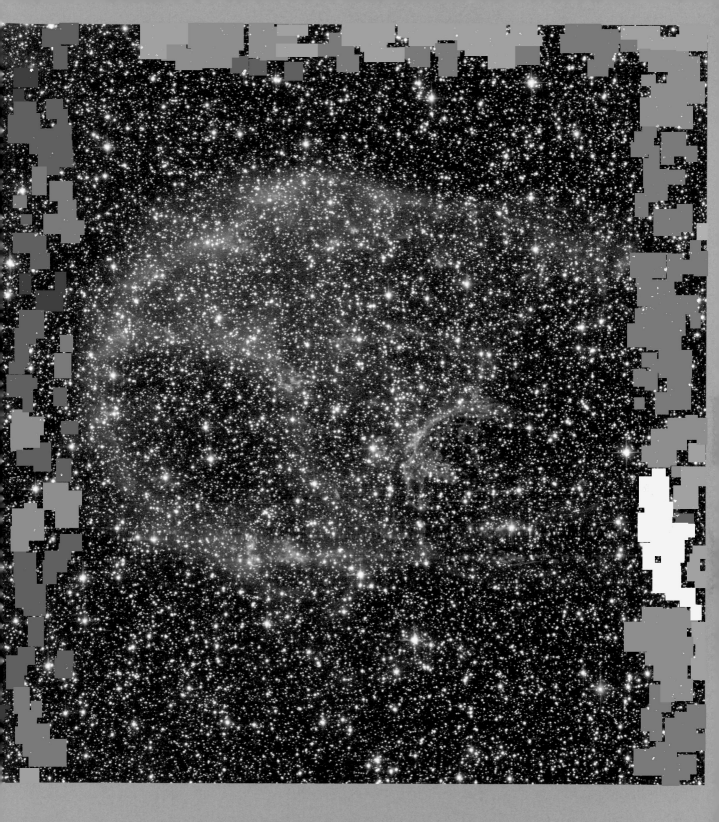

N132D: 160,000 LIGHT-YEARS

The supernova remnant called N132d has evolved into an unusual horseshoe shape, which is seen here against a backdrop of thousands of stars. When a star about twenty times more massive than the Sun collapsed and exploded, it produced shock waves that heated gas around the star to millions of degrees detected in X-rays (blue). Optical data (pink and purple) reveal cooler gas and a small, bright crescent-shaped cloud of gas.

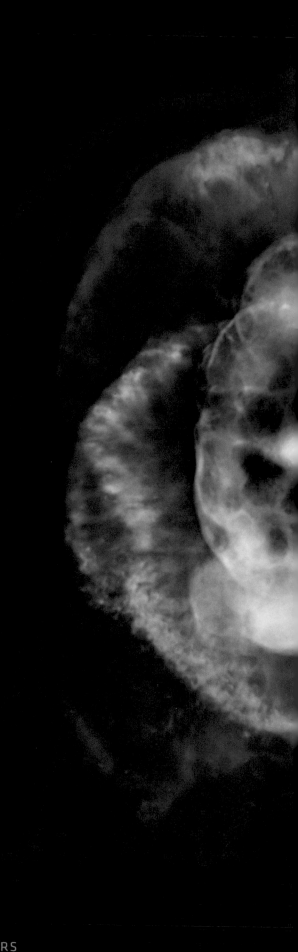

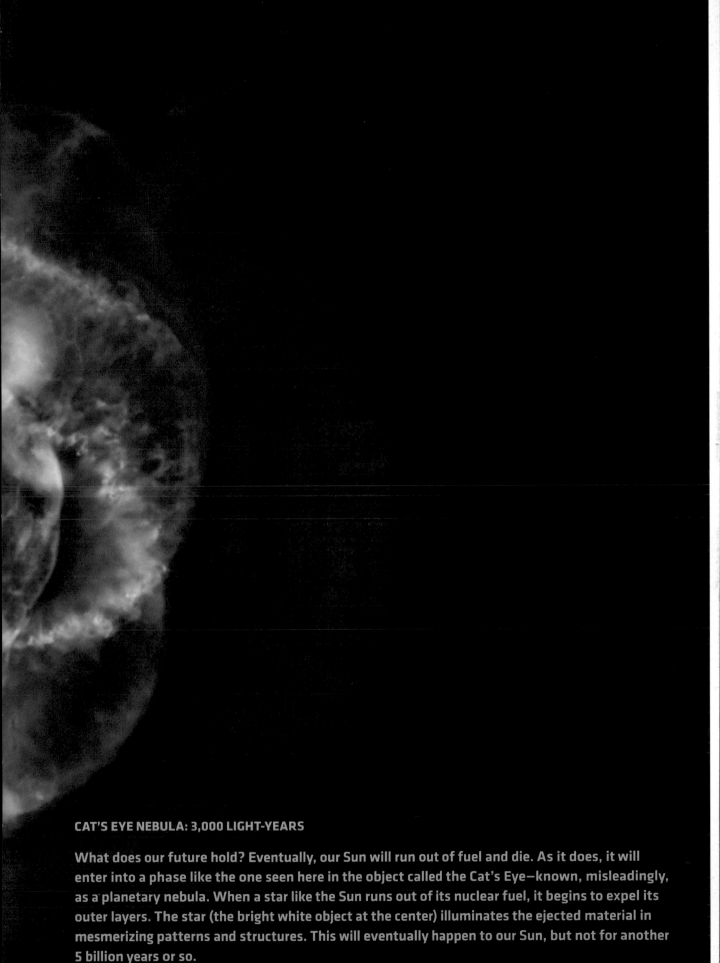

CAT'S EYE NEBULA: 3,000 LIGHT-YEARS

What does our future hold? Eventually, our Sun will run out of fuel and die. As it does, it will enter into a phase like the one seen here in the object called the Cat's Eye—known, misleadingly, as a planetary nebula. When a star like the Sun runs out of its nuclear fuel, it begins to expel its outer layers. The star (the bright white object at the center) illuminates the ejected material in mesmerizing patterns and structures. This will eventually happen to our Sun, but not for another 5 billion years or so.

6

THE MILKY WAY

Why should I feel lonely? Is not our planet in the Milky Way?

— HENRY DAVID THOREAU

I F STARS ARE LIKE PEOPLE, THEN WHERE DO THEY LIVE? Like a majority of humans on Earth, they cluster together in cities. The cosmic equivalent to an urban environment is a galaxy. In the 1920s, an astronomer named Edwin Hubble, who would be recognized decades later by having a now very famous telescope named after him, figured out that there were pockets of stars outside of the Milky Way. He dubbed them "island universes." We know them as galaxies.

For thousands of years before Hubble, humans gazed up at the white band of light that stretches across our night sky. The Greek word for "milk" is *galax*, which gives us the word "galaxy," and the ancient Greeks called the white band overhead the "milky circle," also known as the "milky way." Today we know that the Milky Way is our home galaxy, and that it is made up of billions of individual stars. The white band that we see from our planet is the central region of the Milky Way. We are floating in a single arm of the galaxy, about two-thirds of the way out from the center.

OUR ISLAND UNIVERSE

We live in a galaxy called the Milky Way. But what exactly does that mean? A galaxy contains billions of stars, and our Milky Way is no exception. Many, if not most, of these stars probably have planetary systems of some configuration around them. So our island universe contains billions of residents of stars and planets.

But that's not all. In addition to inhabitants, a cosmic city needs to have infrastructure. This comes in the form of gas and dust, akin to streets and sidewalks that connect the places together. When we think of dust on Earth, our minds may turn to those pesky clumps of dirt and pet hair that gather under pieces of furniture. The dust in the Universe, however, is different than the scourge of vacuum cleaners here on Earth. The term refers to small particles consisting of just a few molecules up to about a tiny fraction an inch up to a few inches (or, one-tenth of one-millionth of a meter). The gas of the Milky Way is mainly hydrogen and helium with traces of other elements thrown in. Both gas and dust are found in hefty quantities in the Galaxy and help to give it its shape. We can trace the Milky Way's gas and dust using different telescopes including those that detect radio waves and infrared emission.

Also essential in the Milky Way are its unseen elements. In a city on Earth, this includes the sewers, pipes, buried cables, and the like. In a galaxy, this takes the form of material known as dark matter. Astronomers gave this stuff such a mysterious name for good reason: It is called "dark matter" because it is dark.

For centuries, astronomers have used some basic laws of physics to measure things that they could not see. Take two objects in space that are separated by a known distance and are in orbit around each other. You know the mass of one of them, but you don't know the mass of the other. If you can tell how fast the object of the unknown mass is moving, you can figure out its mass using a few simple equations. (Try a Web search for "Newton's Laws" for more about them.)

Stretching across this image is part of the Milky Way's plane, where vast clouds of bright stars and clumps of dark dust are visible.

In the 1960s, astronomer Vera Rubin and others tried to play essentially the same game with galaxies. She was looking at how fast clouds were moving around the outer edge of the galaxies, when she got a big surprise. The mass of galaxies was five times greater than what made sense when she added up all of the stars, gas, dust, and everything else you could see with a telescope.

Abell 1689 is a cluster of galaxies a little more than two billion light-years from Earth. Dark matter cannot be seen, but a map of its distribution is visualized here as a blue layer over the normal galaxies (in yellow and white).

Astronomers have since concluded that there is a lot of material that we cannot see, at least with any type of telescope built so far, that fills the galaxies and the rest of the Universe. Astronomers came up with the term "dark matter" because it doesn't give off light. It does have a gravitational effect, however, which explains how it influences how the clouds in the galaxies move. So astronomers know that dark matter is there by looking at its impact on the stuff that we can see with telescopes.

There are several different ideas of what dark matter could be, and many scientists are feverishly working on the problem, using both particle physics laboratories on Earth and telescopes in space. The bottom line is, however, that the true nature of dark matter is just about as mysterious as it was when scientists first detected it decades ago.

MAPPING THE MILKY WAY

Now that we have covered the basic constituents of what's in a galaxy, we can look at the layout of our Milky Way. (As we do with the Moon, we use capitalization to clarify what galaxy we are talking about. We use "Galaxy" to refer to our Milky Way, and "galaxy" when we are referring to any other one.) To give a very loose visual illustration, the Milky Way is shaped like a giant pancake with a softball stuck through its center. Most of the stars in the Milky Way, including our Sun, lie in the Galaxy's disk. The Milky Way's disk spans more than 100,000 light-years across. Like Saturn's rings, the disk of our Galaxy is comparatively very flat—only about a thousand light-years thick. In other words, the Milky Way's disk is about the shape and thickness of a vinyl record or an oversized compact disc.

Rather than a perfect circle, the Milky Way's disk is more of a spiral in shape, with several arms sweeping out of a central hub like a nautilus. Because of the distinct shape, astronomers classify the Milky Way and others like it, not surprisingly, as spiral galaxies. Earth sits on one of the Milky Way's estimated four major spiral arms. In other words, in a city that is in space, we live in galactic suburbia.

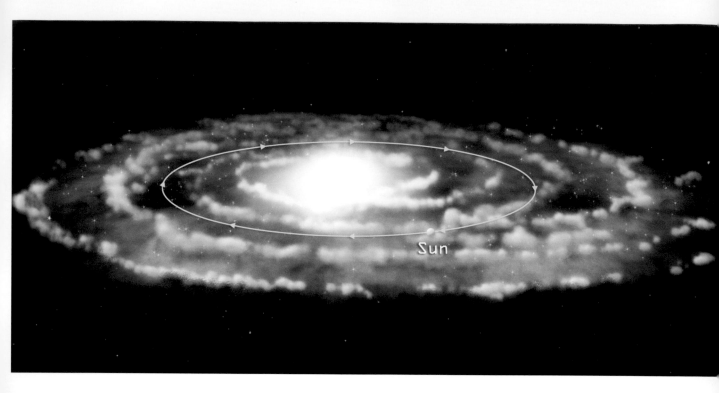

Our Solar System orbits around the Milky Way. At present, we cannot take photos of ourselves from outside of our Galaxy, but scientists have been able to map out what we should look like from a more alien perspective.

THE GALACTIC CENTER

If we head toward the center of the Milky Way, the disk thickens into a ball-like shape where millions of stars are found (think of the softball). At the very center of our Galaxy lives a giant black hole that astronomers call Sagittarius A*, or, in shortened form, Sgr A* (pronounced "saj A-star.") This monster weighs in at some four million times the mass of the Sun.

Scientists think that most galaxies have a giant black hole living at its center. However, Sgr A* is special because it belongs to our Galactic home. Because it is so close to us, we can study it more easily and in more detail than any other black hole like it.

It hasn't always been possible to get much information about Sgr A* and its neighborhood in the center of our Galaxy. Most of the light from it that optical telescopes can detect—which is all we humans had until a handful of decades ago—is absorbed by gas and dust in the disk of the Milky Way. We have to look through 26,000-light-years worth of that gas and dust to see the light coming from the Milky Way's downtown. If we can use only our eyes or even powerful optical telescopes, we won't be able to see very much.

Our understanding of the center of the Galaxy improved dramatically once we developed telescopes that could detect different types of light and penetrate the gas and dust. For example, radio waves are so long that they are not blocked by the Milky Way's gas and dust. Likewise, X-rays are so energetic that they pass through most of the dust and gas as well. Radio telescopes were not developed until the 1930s. Scientists needed to wait until the Space Age in the 1960s when they could begin to launch instruments and detectors sensitive to X-rays and other types of high-energy radiation above the Earth's atmosphere.

Using radio and X-ray telescopes as well as other sources of information, astronomers have been able to watch what the giant black hole at the center of our Galaxy is doing. And what is it that our Milky Way's black hole occupies itself with? For the most part, it likes to sit quietly. As will see in the next chapter, some black holes are very actively ingesting material (we can tell that by how much light is given off around them). Ours, however, apparently doesn't have much to eat, except for the occasional small snack. (By small, we mean something the size of a planet or even a comet.) That doesn't mean that it's not interesting to watch. On the contrary, scientists are able to learn a lot about black holes just by watching what a plain-vanilla black hole like Sgr A* is—and isn't—doing.

Instead of being mere harbingers of destruction, the black hole at the center of the Milky Way and those in other galaxies are probably just as responsible for growing and nurturing. To put it another way, the Milky Way's black hole may be seen as an aggressive downtown industry. It consumes some of its neighbors, but it might also deserve credit for some important positive steps in the region. Not only do black holes ingest material, but they also send out energy in the form of jets that are generated in the disk that surrounds the black hole. This output of energy helps trigger the birth of stars, regulates how quickly the galaxy may form, and contributes to the galaxy's evolution in other useful ways.

TOP TO BOTTOM:

This map shows a panoramic view of the sky in infrared light. Compiled of the locations of almost 100 million stars, most prominent is the dense bulge of stars around the disk of our Milky Way. Across the plane are the dark dust and gas clouds.

This illustration depicts a stellar-mass black hole that comes from the collapse of a massive star. The black hole pulls material from a massive, blue companion star toward it. This material forms a disk (shown in red and orange) that rotates around the black hole before falling into it or being redirected away from the black hole in the form of powerful jets.

In addition to its enormous black hole, the Milky Way's downtown—the Galactic Center—contains many other intriguing destinations. To start, there are many smaller black holes sprinkled throughout the region, which we talked about in Chapter 5. These are the "little" black holes, or stellar black holes, that form from the collapse of giant stars, and we can see them if they are in orbit with a regular star (like the Sun). There are also cores of dead stars (so-called neutron stars), debris from exploded stars (supernova remnants), and giant structures of gas and dust that may be on the precipice of being drawn toward the monster black hole in the middle.

OUR SOLAR SYSTEM ON THE MOVE

Everything in our Galaxy is in a slow orbit around its center—and this includes our Solar System. It will take about 230 million years for us to complete one trip around the Milky Way. In the meantime, our Solar System is bobbing very slightly up and down with respect to the plane of the Galaxy. We are now heading up and out of the plane of the Galaxy, but the gravitational pull of the disk will eventually pull us back toward the disk and through it. The cycle will then repeat many times over for billions of years.

Sun →

Our Milky Way and the Sun's approximate position within it.

GLOBULAR CLUSTERS

There are yet more inhabitants of the Milky Way, ones that belong to its ZIP code even though they exist far above and below the disk of the Galaxy or even its bulge at its center. As we mentioned, stars sometimes live in bunches known as globular clusters. These tightly packed groups of stars may be some of the oldest objects in the Galaxy—up to ten billion years old or more. It's still a mystery why these geriatric stellar homes ended up scattered along the outskirts of the Galaxy.

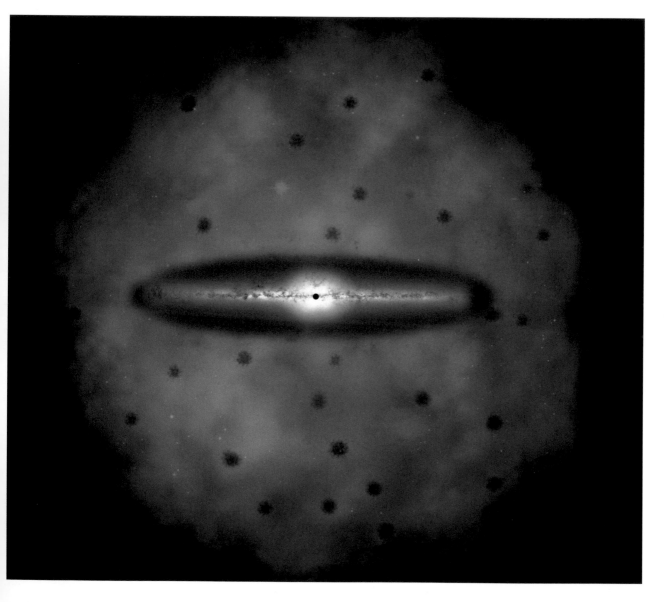

This illustration depicts the pieces that make up our own home Galaxy. This includes the central black hole (black dot at center), stellar disk (white), stellar bulge (red-orange), thick disk (dark purple), globular clusters (red circles), and dark matter halo (gray). The stellar disk is about 100,000 light-years in diameter. A halo of dark matter extends to a diameter of at least 600,000 light-years.

For a long time, astronomers thought that globular clusters marked where the Galaxy ended. Today, however, they know that our galactic sprawl keeps going beyond where these and all other stars stop. The boundaries, however, get harder to mark off once we get to this point. That's because much of the material in the outer reaches of the Milky Way is probably made up mainly of dark matter. Just how far the dark matter reaches—which, as we know, is notoriously hard to detect—and in which directions is a puzzle that astronomers are still trying to unravel.

NEIGHBORING SATELLITES

As we have discussed, our Milky Way is shaped like a messy pancake, with the Earth sitting about two-thirds of the way from the pancake's center. While astronomers have learned much by looking through this thin yet stuffed disk, it can also be helpful to have a different vantage point.

This is where having friends and relatives in other cities really helps. The Milky Way is not alone in its journey through the Universe. Rather, it travels with several smaller satellite galaxies. The most famous of these are the Large Magellanic Cloud (LMC) and the Small Magellanic Cloud (SMC), two galaxies that orbit around the Milky Way. The LMC and SMC were named after the Portuguese navigator, Ferdinand Magellan, who, along with his crew, used the southern sky to explore the world. Fittingly, you can see both the SMC and the LMC without a telescope from the Southern Hemisphere if you know where to look.

The LMC is found about 160,000 light-years from our Milky Way—practically next door in cosmic terms—while the SMC is a few doors down at just around 200,000 light-years away. Unlike the Milky Way, which spans 100,000 light-years across, the LMC is a mere 14,000 light-years wide and the SMC is only 7,000, making them veritable pipsqueaks. However, the locations of the LMC and SMC make them incredibly valuable to astronomers.

Because the LMC and SMC are found almost straight up from the Milky Way's disk, there is little interference from the dust and gas in our Galaxy. And since the Magellanic Clouds are so close, astronomically speaking, scientists can study things like supernova remnants and pockets of star formation with a clear view.

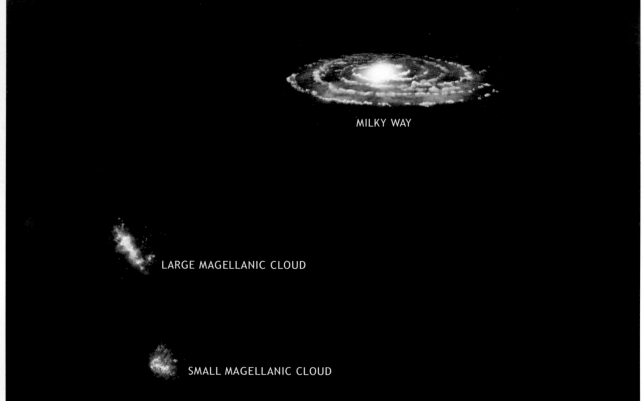

TOP TO BOTTOM:

This photograph shows the southern Milky Way (band of light going up and down on the left), Large Magellanic Cloud (upper right), and Small Magellanic Cloud (lower right), as well as the outline of a cactus (bottom right).

This illustration shows the estimated locations of the Large Magellanic Cloud, the Small Magellanic Cloud, and the Milky Way galaxy.

One other advantage of the LMC and SMC is that we know the distance to these little galaxies so well due to a specific type of star, called Cepheids. These stars are variable, meaning that they change brightness over time. In the early twentieth century, a young woman named Henrietta Leavitt, working at the Harvard College Observatory (she and the other women who worked there at the time were called "computers"), figured out that there was a distinct and consistent pattern for how Cepheids behaved. This meant that if you could find one, you could figure how far away it was. Because the LMC and SMC are close enough to see individual stars, astronomers have been able to determine just how far away these two neighboring galaxies really are.

It is worth noting that techniques such as the one Henrietta Leavitt developed for the Cepheids work only for stars that are relatively nearby and that have certain stable characteristics we can count on. Once we get farther out into the Universe, there are more unknown factors in any distance measurement that scientists try to make.

THE LOCAL GROUP

Another special member of the Milky Way's regional neighborhood is the Andromeda Galaxy. Known officially by astronomers as Messier 31, Andromeda is our Galaxy's sister in many ways. Like the Milky Way, Andromeda is a spiral galaxy. You can spot it within the constellation of Andromeda, which is where the galaxy gets its name.

Unlike the SMC and LMC, Andromeda is roughly the same size as the Milky Way. At a distance of 2.5 million light-years, Andromeda provides a chance for us to look at ourselves in the galactic mirror, so to speak. One of the most intriguing facts about Andromeda is that it will collide with our own Milky Way several billion years from now.

This is a cosmic good news/bad news situation. Let's start with the "bad" news: There is nothing the Milky Way can do to avoid this collision. The "good" news, however, is pretty good. First, this galactic encounter won't begin for several billion years.

Second, most of the stars and planets (including Earth, if we're still around) probably won't be affected. Even though the galaxies will merge, there is so much room in between the stars that many will pass by each other unbothered. One cool thing about this eventual collision with Andromeda: the view of the night sky for anyone living on a planet in either galaxy would look completely different during and after the merger.

Finally, the Milky Way, Andromeda, the LMC, and the SMC, all belong to what astronomers refer to as the Local Group. It sounds like a neighborhood clique—and in some ways it is. However, what holds this group of more than thirty galaxies together isn't a random hobby or social standard. Instead, the galaxies of the Local Group are held together by gravity. We will look closer at how galaxies rarely travel alone in the upcoming chapters.

Will this be how our Galaxy looks when it collides with Andromeda? This image from Hubble shows a pair of galaxies about 300 million light-years from Earth that will eventually merge into one. This cosmic pairing known as NGC 4676 has been nicknamed "The Mice" because of the long tails of stars and gas stretching out from each galaxy.

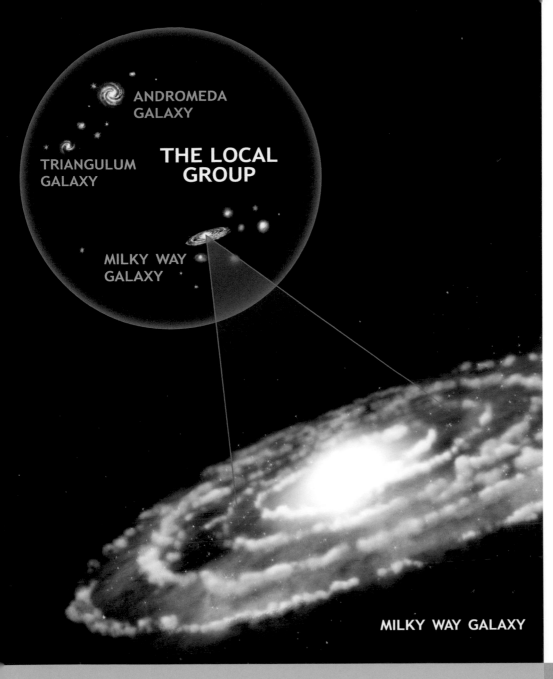

ANDROMEDA
GALAXY

TRIANGULUM
GALAXY

**THE LOCAL
GROUP**

MILKY WAY
GALAXY

MILKY WAY GALAXY

*Our Galaxy is not alone
in the universe,
but congregates in its
neighborhood of
galaxies called the Local
Group, as illustrated
here. Our Local Group
has over 30 other known
galaxies, including
Andromeda.*

OUR SUN IS ONE OF BILLIONS OF STARS IN THE MILKY WAY,
OUR HOME GALAXY.

THE MILKY WAY CONTAINS A HUGE THIN DISK,
AND THE SOLAR SYSTEM SITS ABOUT TWO-THIRDS OF THE
WAY OUT IN IT.

AT THE CENTER OF OUR MILKY WAY IS A GIANT BLACK HOLE—
BUT IT POSES NO DANGER TO US ON EARTH.

SINKING YOUR TEETH INTO THE MILKY WAY

THE MILKY WAY

One of the most inspiring naked-eye objects in the sky is our own Galaxy, the Milky Way. This mosaic of photographs—taken from the ground in Germany and Namibia—shows a very wide field view of our Galaxy. You can see some of the billions of stars that make up our galactic neighborhood in this image, along with reddish clouds of gas and a dark obscuring lane of dust that stretches across the plane of our Galaxy.

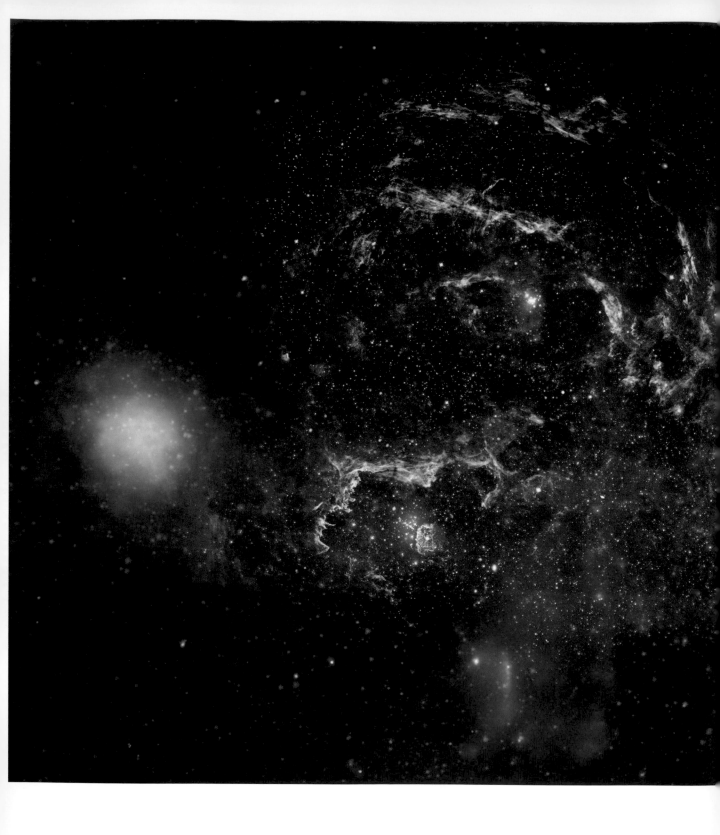

THE CENTER OF THE MILKY WAY: 26,000 LIGHT-YEARS

X-ray, optical, and infrared data from three of NASA's telescopes in space have been combined to produce this stunning image of the central region of the Milky Way. One slice of infrared light from Hubble (yellow) outlines energetic regions where stars are being born. Other infrared data from Spitzer (red) show glowing clouds of dust containing complex structures. X-rays from Chandra (blue and violet) reveal gas heated to millions of degrees by stars that have exploded and material flowing away from the Galaxy's supermassive black hole.

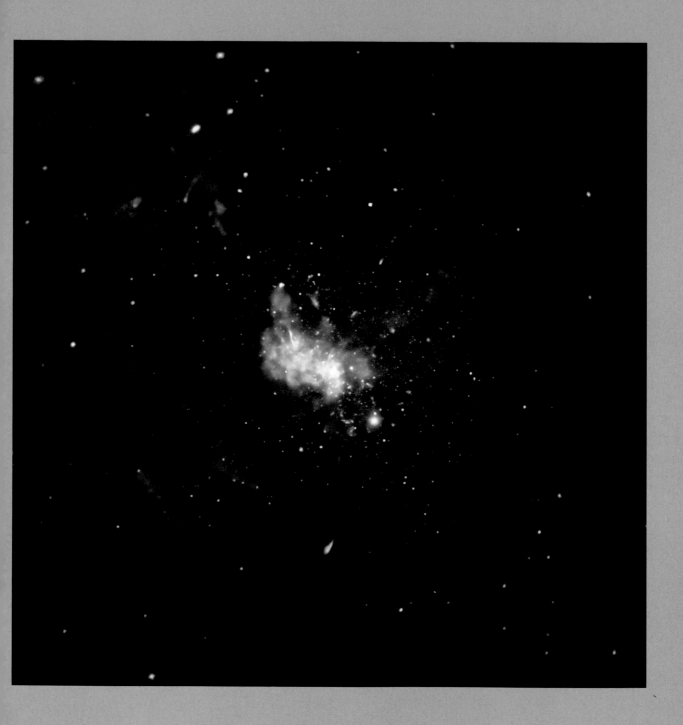

SAGITTARIUS A*: 26,000 LIGHT-YEARS

This Chandra X-ray image of Sagittarius A*, the Milky Way's giant black hole, and the region around it was based on almost two weeks of observing time. Using these data, scientists have developed ideas as to why Sagittarius A* (buried in the white area in the center of this image) seems to ingest so little material compared to similar black holes found in other galaxies. Outward pressure in the region around Sagittarius A* may be causing the black hole's fuel supply of gas and dust and stars to move away instead of heading in to be consumed.

THE MILKY WAY GALAXY'S ANTICENTER: 6,000–12,000 LIGHT-YEARS

The so-called interstellar medium, simply put, is the material found in between stars. This image shows the complexity of the interstellar medium in the direction directly away from the center of our Galaxy (called the "anticenter"). Dust, which gives off infrared radiation, has been colored blue while radio-emitting regions are pink. Supernova remnants appear as red and yellowish bubbles. On the far left of the image, regions of pink and blue are dusty nebulas that are also forming stars.

THE HEART OF HERCULES GLOBULAR CLUSTER: 25,000 LIGHT-YEARS

This image, taken by Hubble, shows the core of globular cluster Messier 13 and provides a clear view of the hundreds of thousands of stars in the cluster, one of the brightest and best-studied clusters known in the sky. The cluster, which is 25,000 light-years from Earth and 145 light-years across, lies in the constellation of Hercules. It is so bright that under the right conditions it is visible without a telescope

OMEGA CENTAURI: 17,300 LIGHT-YEARS

A million lights fill this view across the core of the globular cluster Omega Centauri, picked out in detail by the Hubble Space Telescope. There are about two hundred of these clusters in our Galaxy, each containing millions of very old stars clumped together into a ball by gravity.

NGC 6752: 14,000 LIGHT-YEARS

NGC 6752 is a brilliant globular cluster deep in the southern sky within the constellation of Pavo, "the peacock." It is a little brighter than the much better known northern globular cluster M13. The colors of the stars are subtle, but this image clearly shows the cool but bright red giant stars whose light dominates the cluster.

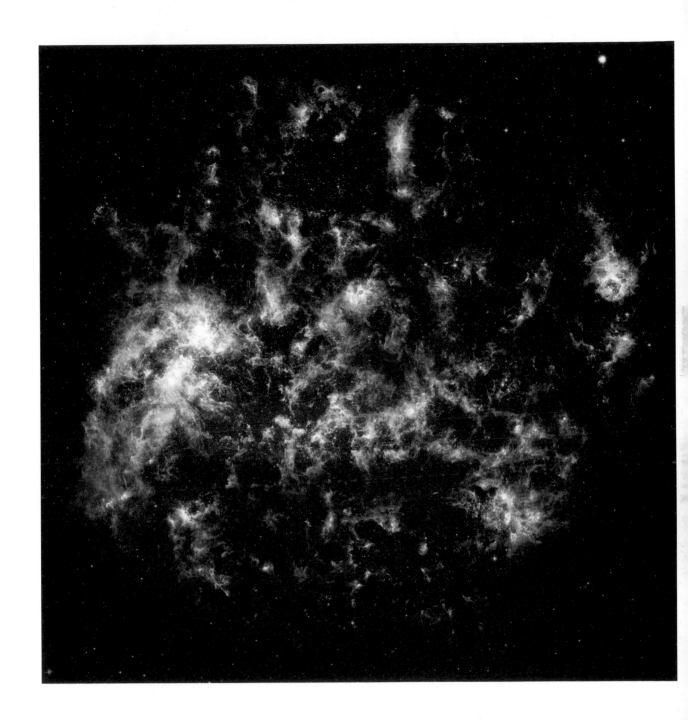

LARGE MAGELLANIC CLOUD (INFRARED): 160,000 LIGHT-YEARS

The Large Magellanic Cloud, the Milky Way's neighbor, is visible in the southern sky as a pale cloudlike object. This infrared image from the orbiting Spitzer Space Telescope reveals huge swathes of hot gas and dust, showing that the Large Magellanic Cloud is a hotbed of star formation.

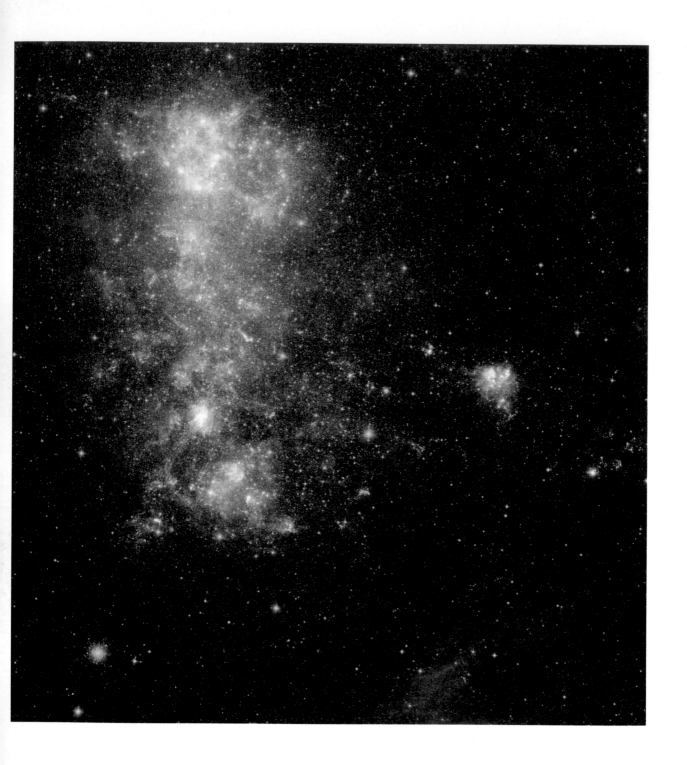

SMALL MAGELLANIC CLOUD (INFRARED): 200,000 LIGHT-YEARS

This infrared view of the Small Magellanic Cloud, taken by NASA's Spitzer Space Telescope, shows the stars and dust in this nearby satellite to the Milky Way. The image shows old stars (blue) and young stars lighting up their young dust (green and red).

SMALL MAGELLANIC CLOUD STARS: 200,000 LIGHT YEARS

Hubble's excellent sharp vision plucked out an underlying population of embryonic stars embedded in the nebula NGC 346 that are still forming from gravitationally collapsing gas clouds. Although they will become full-fledged stars, they have not yet ignited their hydrogen fuel to sustain nuclear fusion. The smallest of these infant stars is only half the mass of our Sun.

ANDROMEDA: 2.5 MILLION LIGHT-YEARS

The closest spiral galaxy to us, the Andromeda Galaxy (also called Messier 31), would appear eight times the size of the full Moon in the sky if our eyes were sensitive enough to see it. Luckily, we can use telescopes to capture Andromeda in its full glory. Spanning 150,000 light-years, Andromeda has a shape very much like the Milky Way with older, yellow stars in the center and younger, blue stars in the spiral arms.

7

GALAXIES BEYOND OUR OWN

I have loved the stars too fondly to be fearful of the night.

— SARAH WILLIAMS

WE THINK THE MILKY WAY IS SPECIAL BECAUSE IT'S OUR very own galaxy. But is it typical in comparison to other galaxies out there?

The short answer is: Yes and no. While there are other galaxies that resemble our Milky Way, there are also billions of others that do not.

Astronomers categorize galaxies mainly based on their shape. This is a fairly good place to start, since most of us started recognizing shapes at a very young age. The types are even called by their shapes: spirals, ellipticals, and irregulars (okay, "irregular" is not a shape, but we think the name is fairly self-explanatory).

Bright knots of glowing gas light up the arms of spiral galaxy Messier 74 (M74). M74 is just slightly smaller than our own spiral galaxy.

The Milky Way is a spiral galaxy, as we have noted. If there is an iconic image of a galaxy, this would be it. Spiral galaxies appear in everything from science fiction movies to cartoons to art posters.

A spiral is defined as a "curve that emanates from a central point getting progressively farther away as it revolves away from the point." Spiral galaxies aren't just featureless pancake-like objects. They have spiral arms. Most of the gas and dust are found in these galactic tentacles. So is Earth, which sits in what astronomers have dubbed the Sagittarius arm of the Milky Way.

Most spiral galaxies have a central bulge of stars in their centers with a giant black hole cloaked inside, just like the Milky Way does. Rather than being some sort of menace, these central black holes may very well be responsible for how big the galaxy gets, how many stars form in it, and a slew of other characteristics. In other words, they act more like zoning boards for the galaxy than neighborhood bullies.

NGC 4565, also called the Needle Galaxy, is a picturesque example of an edge-on spiral galaxy. Located about 30 million light-years from Earth, it has a bright central bulge that juts out above the dust lanes.

Elliptical galaxy ESO 325-G004 is as massive as 100 billion suns. There are also thousands of globular star clusters circling the galaxy.

How did spirals get their shape? That's still a big question, but many astronomers think they built up over time through mergers of smaller clumps of stars. Eventually, the spiral-galaxy-to-be started to spin as these ongoing collisions revved it up. Eventually, the galaxy flattened out and the spiral arms developed.

As an aside, we think all spiral galaxies are turning—including completely mature galaxies like the Milky Way—but very, very slowly. Scientists estimate that it takes the Milky Way about 230 million years to make one complete rotation.

ELLIPTICALS

The second major type of galaxies is the elliptical. As the name implies, these galaxies are shaped like an egg or a ball. They do not have the majestic arms of a spiral, nor do they have a telltale bulge at their center.

Ellipticals contain lots of old stars and very few young ones. In fact, they don't have much gas and dust in between their stars, as spirals do. This means that it's harder for enough gas to condense to form new stars. Scientists think that ellipticals are actually the results of mergers or collisions between other galaxies, including spirals. These collisions were much more frequent when the Universe was younger, so most of the ellipticals are typically much older than the spiral galaxies we have talked about. Elliptical galaxies may look a little dull, but don't let that mislead you into missing just how important these objects are.

A GALAXY BEHAVING BADLY

Sometimes galaxies do not keep to themselves. They act in ways that can impact—quite literally—their environment and others around them. Take the galaxy known as 3C321, which astronomers have nicknamed the "Death Star" galaxy. If you are familiar with the original *Star Wars* trilogy, you'll remember that the Death Star is a giant weapon that can destroy good-guy planets with a bad-guy superlaser beam.

This galaxy is not quite so calculating, but the black hole in its center is generating an energetic beam that is slamming into another nearby galaxy. We don't know exactly what effect this is having on the unsuspecting neighbor, but it's probably not a good one. Jets from black holes like the one in the Death Star galaxy produce large amounts of gamma rays and X-rays, which are lethal to any life that we know of in large doses. Let's just hope planets in the galaxy getting pummeled have a way to get in touch with Yoda.

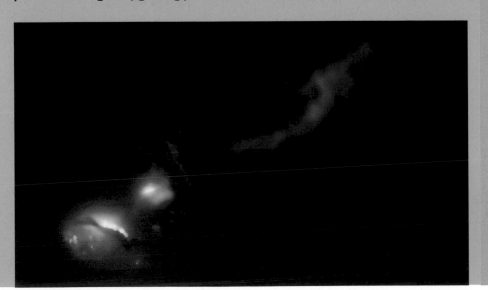

3C321 is located about 1.4 billion light-years from Earth. Four types of light have been combined in this image, including X-ray (purple), optical (red), ultraviolet (orange), and radio (blue) data.

IRREGULARS

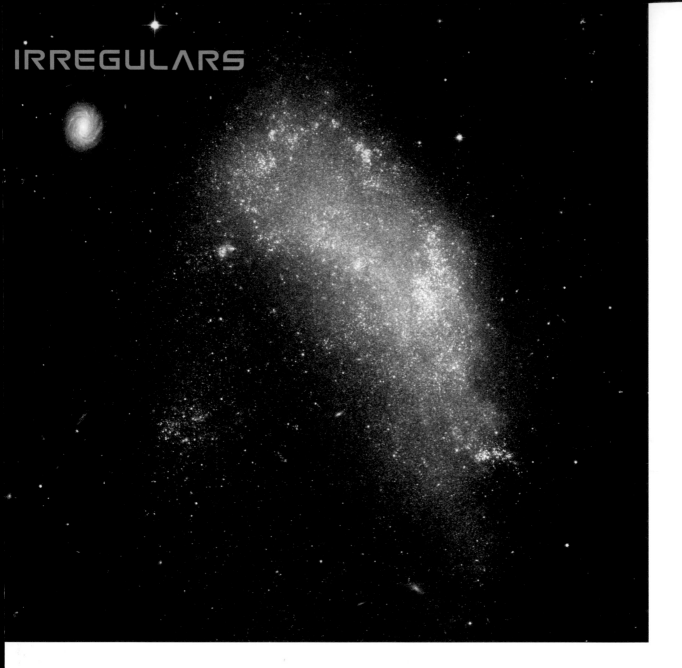

NGC 1427A is an example of an irregular galaxy about 62 million light-years from Earth. This optical image shows many hot, blue stars that have formed quite recently, indicating that a lot of star formation is happening in this irregular galaxy.

The third kind of galaxy is called "irregular." As you may have guessed, irregular galaxies do not have easy-to-identify shapes. Instead, they are messy. Some may have been the victim of a hit-and-run from another galaxy. Others might have been distorted by being too close to another very large galaxy whose gravitational effects were huge. Irregular galaxies, which can be different shapes, sizes and ages, are thought to make up about 25 percent of all galaxies. Think of irregulars as the miscellaneous bin at your favorite department store.

GENUS, SPECIES, FAMILY

These three categories—spiral, elliptical, and irregular—cover the major bases as far as galaxies are concerned. If you are particularly interested in how galaxies are classified, however, you will find many, many more subtypes if you look.

That's because astronomers, like biologists, love to classify things. Do you remember all of those levels of classification in biology—genus, species, family, and so on? Well, astronomers have developed many levels of subcategories for galaxies, not to mention lots of other stuff they find. Scientists do these very detailed groupings because they are trying to find patterns and relationships in the vast amounts of information they encounter.

In other words, for every neat spiral or elliptical galaxy that we see, there are others that defy what scientists were expecting of them. That is one of the things that can make science so fun: You never know when you are going to find something that will make you rethink your ideas and potentially come up with new ones.

GALACTIC FACTS

OUR MILKY WAY GALAXY IS AN EXAMPLE OF A SPIRAL GALAXY, BUT GALAXIES COME IN MANY DIFFERENT SHAPES AND SIZES.

WINDS FROM SOME GALAXIES ARE HELPING TO SEED THE NEXT GENERATION OF STARS AND PLANETS WITH THE SAME KINDS OF MATERIAL THAT WE NEEDED TO GET GOING HERE ON EARTH.

The nearby dwarf galaxy NGC 1569 is a hotbed of stellar birth that blows huge bubbles in the main body of the galaxy.

GALACTIC WINDS

Galaxies are not just static, unchanging islands of gas, dust, and stars. On the contrary, some galaxies are incredibly active.

Take, for example, the galaxies that astronomers call "starbursts." These are galaxies that are undergoing a baby boom of stars. These galaxies are birthing stars at a rate of tens or even hundreds of times that of a "normal" galaxy. Along with this activity on the stellar birth scale, these galaxies are also weathering blistering winds flowing from their centers.

These galactic superwinds, mainly due to the powerful radiation from the frenetic newborn stars and other older stars that have exploded as supernovas, can be seen stretching all of the way across the galaxy.

M82 is a starburst galaxy with a superwind. It is also called the Cigar Galaxy because of its long, thin shape.

Similar to a wind on Earth, these galactic winds are important because they move particles from one place to another. Astronomers have found evidence that these superwinds contain things like carbon, nitrogen, oxygen, and iron—in other words, the stuff necessary for life as we know it. So these galactic winds are helping to seed the next generation of stars and planets with the very material that we needed for life to get going here on Earth.

Quasars are the central regions of some galaxies where a black hole is on an eating extravaganza. Because they have an unusual amount of stuff—mainly gas and dust—to ingest, the black holes in quasars are extremely active (think of a black hole at an all-you-can-eat buffet). This means that they shine brightly in many different types of light, making quasars some of the brightest objects in the entire Universe.

3C 273 is a quasar located about 3 billion light-years from Earth, and was the first quasar ever to be identified.

Indeed, while we have made it seem easy to sort galaxies into the three main categories—spiral, elliptical, and irregular—it's often not that simple. We live in an age where telescopes and observatories are gathering incredible amounts of data. Projects like the Sloan Digital Sky Survey (a telescope in New Mexico that is continuously observing the sky) are taking millions of images of the cosmos, and finding galaxies that are billions of light-years away.

Because modern telescopes are powerful, many of these very distant galaxy images are difficult to interpret—especially for a computer—because they are mere smudges. Rather than being able to rely solely on automated computer programs to correctly classify the thousands and thousands of galaxies in their databases, astronomers have enlisted the broader public to help. Through a project called Galaxy Zoo, they have invited anyone who is interested to take a brief tutorial and start helping to classify these galaxies.

A catalog of just some of the many galaxies found by participants of Galaxy Zoo, composited together into one montage.

Since its launch in 2007, the Galaxy Zoo project has not only enabled the correct classification of hundreds of thousands of galaxies, but it has also led to new and exciting research results about the details of these galaxies. The Galaxy Zoo is an example of what is called "citizen science," a recently dubbed term for scientific research that nonscientists can participate in and make significant contributions to. If you are interested in getting involved, visit the Galaxy Zoo website at www.galaxyzoo.org.

COLLIDING AND
INTERACTING GALAXIES

As we mentioned in the previous chapter, the Milky Way is on a collision course with our neighboring galaxy Andromeda, an event set to happen several billion years from now. While this sounds like a big deal (and it could be for anyone or anything living in those galaxies in 5 billion years), this collision is not as rare as you might think.

Galaxies collide often. These galaxy collisions are not just pieces of cosmic violence to gawk at; rather, they have been and continue to be important for shaping our present-day Universe and how it looks today. We see galaxies colliding throughout the Universe, but these events happened even more frequently in the distant past, when galaxies were much closer together.

We have learned a lot since Edwin Hubble called them island universes back in the 1920s, but there is still so much we want to know about galaxies. For example, how exactly do galaxies form, and how do they evolve over time? As you read this, scientists around the world are trying to tackle these questions. Others are trying to figure out the role that giant black holes play in how galaxies grow up. It's a chicken-and-egg sort of question: Do the black holes form first, or do the galaxies? Or do they grow up together as a codependent pair?

We have learned a lot since Edwin Hubble called them island universes back in the 1920s, but there is still so much we want to know about galaxies. For example, how exactly do galaxies form, and how do they evolve over time? As you read this, scientists around the world are trying to tackle these questions. Others are trying to figure out the role that giant black holes play in how galaxies grow up. It's a chicken-and-egg sort of question: Do the black holes form first, or do the galaxies? Or do they grow up together as a codependent pair?

In the next chapter, we'll get into some of the things we have found out about the Universe when we study galaxies not as individuals, but as members of the groups and supergroups they tend to hang around in.

NGC 4945: 13 MILLION LIGHT-YEARS

The spiral galaxy NGC 4945 is a close neighbor of the Milky Way. Resembling our own Galaxy, NGC 4945 also hides a supermassive black hole behind its thick, ring-shaped structure of dust. But unlike the black hole at the center of our Milky Way, which is very quiet, the black hole at the center of NGC 4945 is consuming surrounding matter at a frantic pace and is releasing tremendous amounts of energy in exchange.

M106: 23.5 MILLION LIGHT-YEARS

M106 is an unusual spiral galaxy located in the constellation Canes Venatici (the Hunting Dogs). Two spiral arms, dominated by young, bright stars, are seen here in visible light, but X-ray and radio imaging (not shown) reveals two additional "anomalous" arms. These unexpected arms consist mainly of gas that is being heated by shock waves.

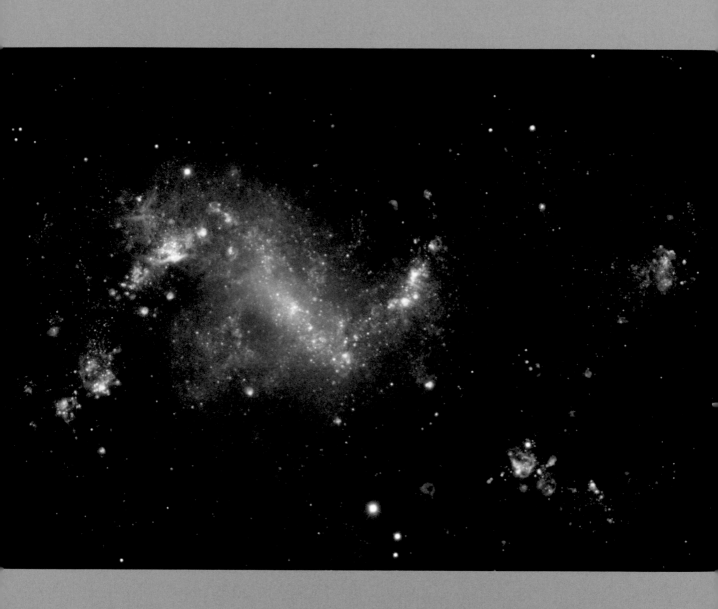

NGC 1313: 14 MILLION LIGHT-YEARS

NGC 1313 is a relatively nearby spiral galaxy. The clouds of hot gas are bubbles, shock fronts, and supernovas formed from the furious star formation going on inside the galaxy. Big bursts of star formation in a galaxy are often triggered by a close encounter with another galaxy, but NGC 1313 is alone, leaving questions as to why it is currently forming stars so rapidly.

M82: 12 MILLION LIGHT-YEARS

The irregular galaxy M82 is seen here in three different types of light: optical (green), infrared (red), and X-ray (blue). M82 is one of the most famous starburst galaxies, where stars are being born at a rate at least ten times faster than in our Milky Way. A close encounter with a neighboring galaxy about 100 million years ago that stirred up the gas and dust is probably responsible for this veritable baby boom of stars.

NGC 4696: 150 MILLION LIGHT-YEARS

NGC 4696 is a large elliptical galaxy. This composite image shows a vast cloud of hot gas in X-rays (red), surrounding high-energy bubbles seen in radio data (blue) on either side of the bright white area, which is where a supermassive black hole sits. Scientists can calculate the rate at which gas is falling toward the galaxy's supermassive black hole and how power is needed to produce the bubbles, which are each about 10,000 light-years in diameter. This tells us how efficiently black holes are able to convert the energy of infalling matter into the energy carried away by jets.

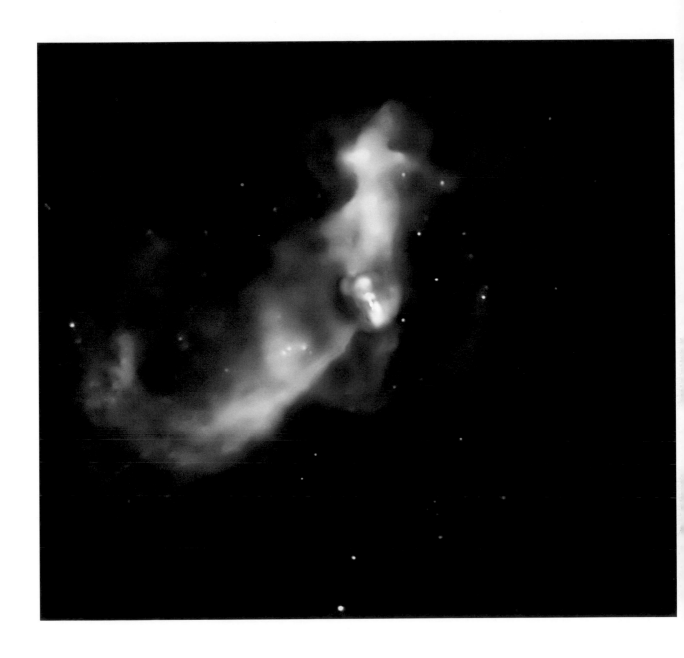

M87: 50 MILLION LIGHT-YEARS

What does a black hole have in common with a volcano? This is not the start of a bad science joke. Instead, it is a legitimate question when you look at M87, a massive galaxy. X-rays from Chandra (blue) show that the cluster surrounding M87 is filled with hot gas. As this gas cools, it falls toward the galaxy's center, where it continues to cool even faster and form new stars. Radio observations (red and orange) suggest that jets of very energetic particles produced by the black hole interrupt this process. This type of interaction actually mirrors many properties of the Icelandic volcano that erupted in the spring of 2010 and disrupted air travel all over the world.

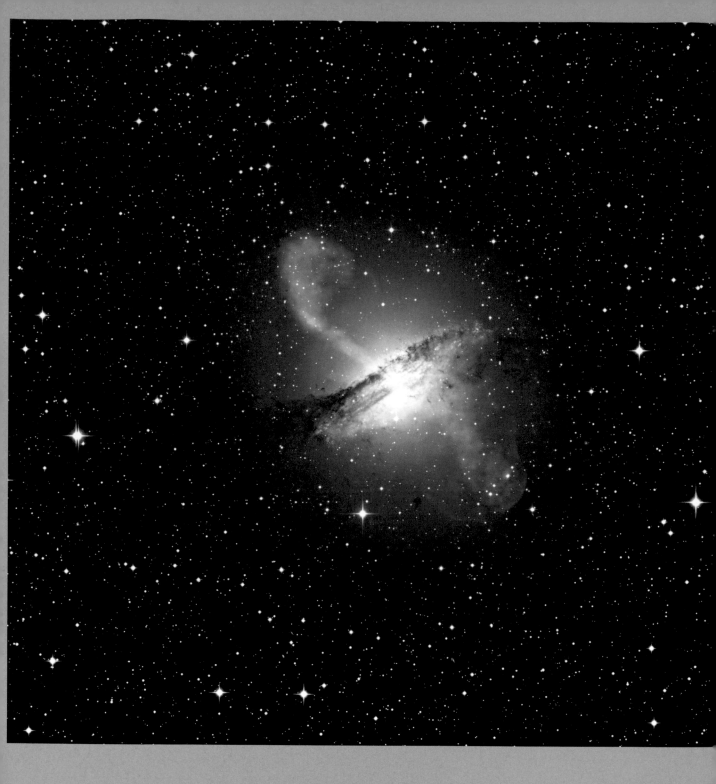

CENTAURUS A: 11 MILLION LIGHT-YEARS

This image of Centaurus A shows a magnificent view of a supermassive black hole's power. Jets and lobes powered by the central black hole in this nearby galaxy are revealed by submillimeter data (orange) from a telescope in Chile and X-ray data (blue) from the Chandra X-ray Observatory. Visible light data from another telescope in Chile shows the dust lane in the galaxy and background stars. The X-ray jet in the upper left extends for about 13,000 light-years away from the black hole. The submillimeter data indicates that material in the jet is traveling with the remarkable velocity of about half the speed of light.

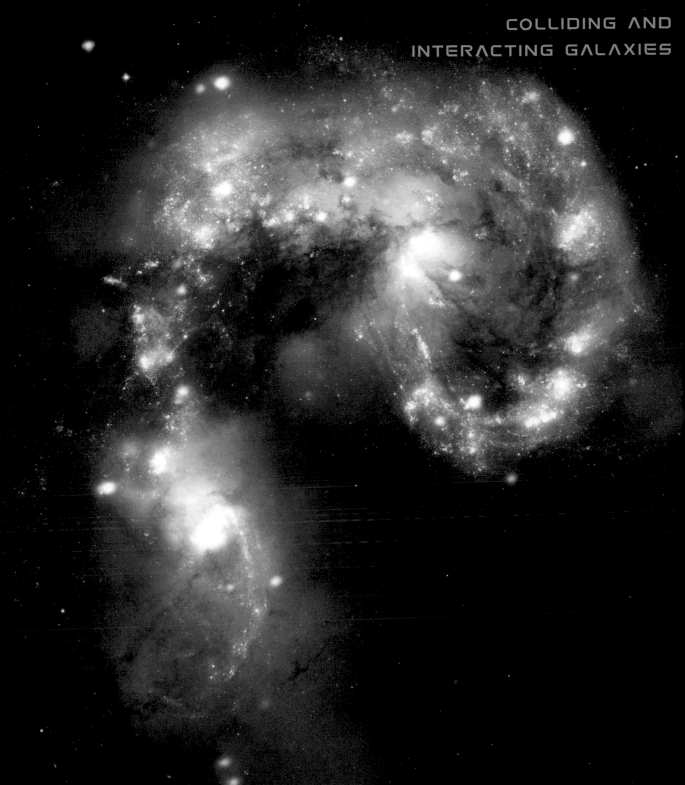

ANTENNAE GALAXIES: 62 MILLION LIGHT-YEARS

The Antennae Galaxies are a pair of colliding galaxies. This composite image contains X-rays from Chandra (blue), optical data from Hubble (gold and brown), and infrared data from Spitzer (red). The X-ray image shows huge clouds of hot gas between the stars that have been injected with rich deposits of elements from supernova explosions. This enriched gas, which includes elements such as oxygen, iron, magnesium, and silicon, will be incorporated into new generations of stars and planets. The bright, pointlike sources in the image are produced by material falling onto black holes and neutron stars that are remnants of the massive stars.

WHIRLPOOL GALAXY: 31 MILLION LIGHT-YEARS

A beautiful face-on spiral galaxy, this optical view of the Whirlpool Galaxy (also known as Messier 51) shows what the Hubble Space Telescope saw when it observed this classic spiral. The red glow depicts enormous clouds of hydrogen gas. Also seen in this image is how the larger Whirlpool galaxy is interacting with its much smaller neighbor, the yellow-colored NGC 5195 on the top left.

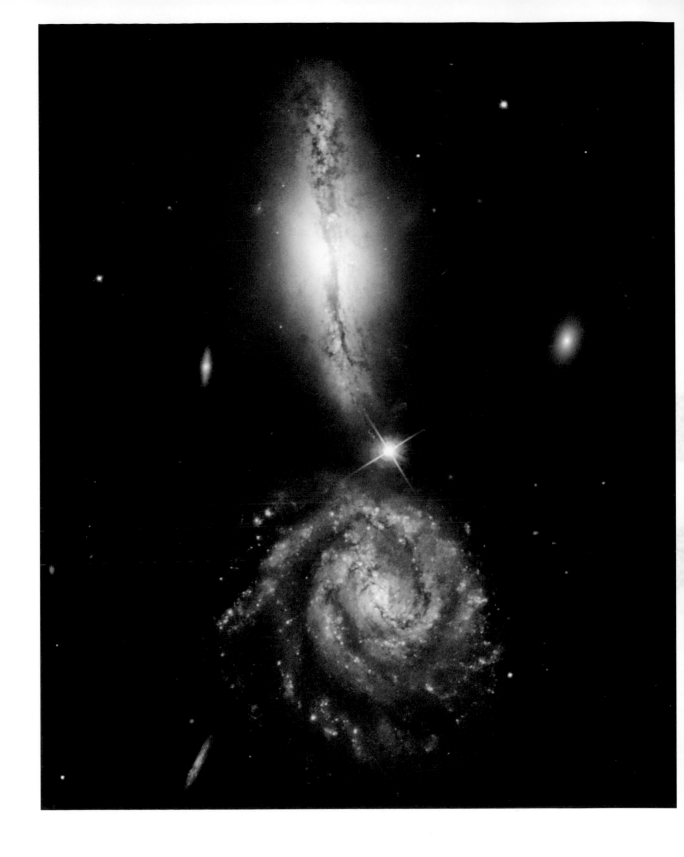

VV 340: 450 MILLION LIGHT-YEARS

It's not every day that the Universe gives us a sign, but this punctuation mark is the exception. This image actually contains a pair of galaxies known as VV 340. X-ray data from Chandra (purple) and optical data from Hubble (red, green, blue) have been combined to show these two galaxies that are in the early stage of an interaction. Millions of years from now, the two galaxies will merge.

NGC 1316: 70 MILLION LIGHT-YEARS

How did this galaxy form? Astronomers turn detectives when trying to figure out the cause of unusual jumbles of stars, gas, and dust like in NGC 1316. At first glance, NGC 1316 looks like an enormous elliptical galaxy, but one that includes dark dust lanes usually found in a spiral. Since there are fewer low mass globular clusters of stars toward NGC 1316's center, scientists suspect that this galaxy had a collision or merger with another galaxy in the past few billion years.

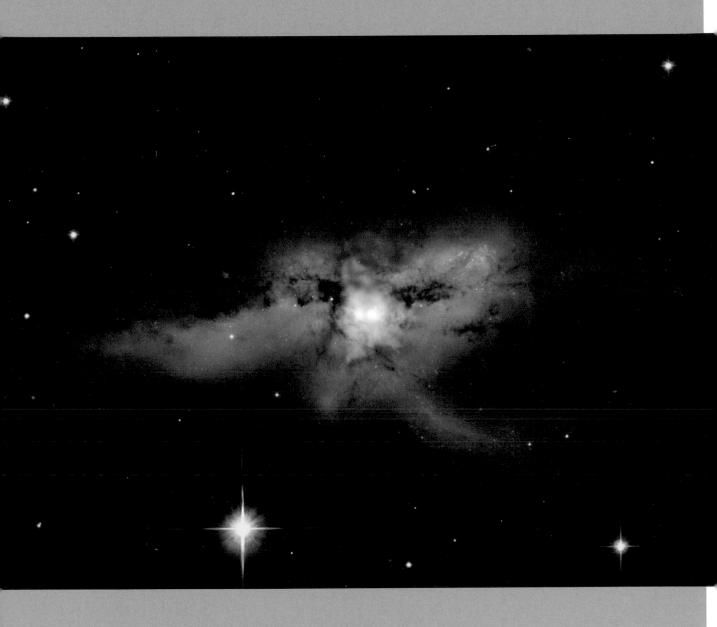

NGC 6240: 330 MILLION LIGHT-YEARS

When two galaxies collide, what happens to their supermassive black holes at their centers? In some cases, the two black holes spiral around one another until they eventually merge as well. In NGC 6240, two supermassive black holes (the bright pointlike sources in the center of this image) are a mere 3,000 light-years apart. Scientists estimate that they have been closing in on each other for about 30 million years and will ultimately merge into a larger black hole millions of years from now.

8

CLUSTERS
OF GALAXIES

In three words I can sum up everything I've learned about life: it goes on.

— ROBERT FROST

Many galaxies are social and form into groups, just as stars do, under the influence of gravity. This group of four galaxies is the brightest in the Hickson catalog (the catalog is named after the Canadian astronomer who created the list in the 1980s). Hickson Compact Group 44, about 60 million light-years away, can be seen through a good amateur telescope.

WE HAVE COME QUITE A LONG WAY FROM OUR relatively small but amazing planet

across the Solar System, through our Milky Way galaxy, and onto the dazzling array of galaxies across the Universe. We have jumped through enormous differences of scale in both size of these objects and the distances between them. But hold on for another big leap. Because as enormous as galaxies are, they are, in fact, just small pieces of some of the largest objects in the Universe: galaxy clusters.

Why do we care about galaxy clusters? After all, we are incredibly far from home here. As it turns out, astronomers can use galaxy clusters to probe some of the biggest mysteries in the entire Universe—dark matter and dark energy. Figuring these two unsolved puzzles out will let us answer the questions that all kids have asked their parents from the back seat: Where are we going, and when are we going to get there?

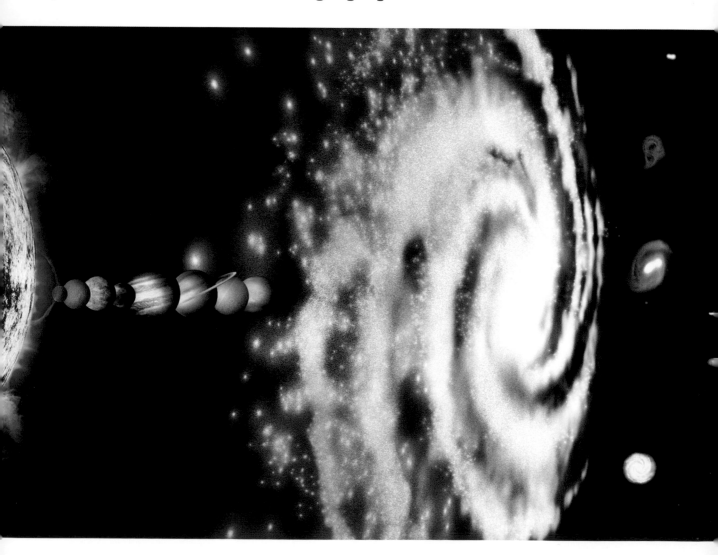

First things first: Let's look at why galaxies are often in clusters. Galaxies, it turns out, like to be around other galaxies. Remember, everything that has mass—whether on Earth or in space—has a gravitational pull. The important thing to keep in mind is the bigger the object, the greater its gravitational power. This leads to galaxies becoming bound to other galaxies in either groups (which have up to about a hundred galaxies) or clusters (which can contains thousands of galaxies.)

As we mentioned in Chapter 6, the Milky Way is indeed part of its own little galactic posse, which astronomers call the Local Group. On even larger scales, the Milky Way and its close neighbors are part of the much larger collection known as the Virgo Super Cluster.

This illustration of our journey through the Universe (not to scale), starts on the left with our Sun and Solar System. Then we move out from our Milky Way galaxy to other local galaxies, and beyond to the most distant parts of the early Universe.

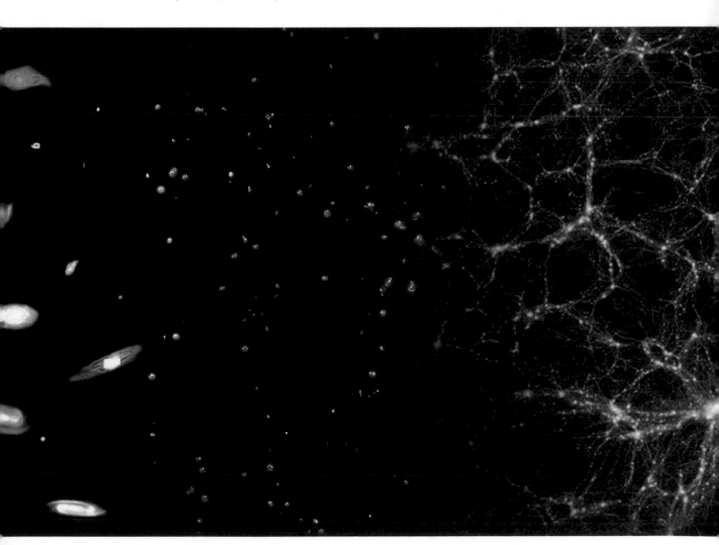

In the foreground is the largest galaxy visible in the image, M81. M81 is gravitationally interacting with M82, just below it with the halo of filamentary red-glowing gas. Many other galaxies are seen in this image, which shows just a small part of the Virgo Cluster of galaxies. Note that this zoo of galaxies is seen through the faint glow of nebulous material in our Milky Way.

MAKING A GALAXY CLUSTER COCKTAIL

Galaxy clusters are made up of about three distinct parts: the individual galaxies, hot gas that fills the space between the galaxies, and dark matter. This is more than a matter of cosmic housekeeping. Each of these constituents is important for different reasons.

First, the individual galaxies hold all of the stars, gas, and dust in these galaxy clusters. While that sounds like a lot of stuff given that we're talking about entire galaxies here, the hot gas is just as

The galaxies are apparent in this optical image of galaxy cluster Abell 520. The blue-colored image pinpoints the location of most of the mass in the cluster, which is dominated by dark matter. Dark matter is an invisible substance that makes up most of the universe's mass. The dark matter map was derived from the Hubble observations by detecting how light from distant objects is distorted by the cluster galaxies, an effect called gravitational lensing. X-ray data, in green, shows the hot gas in the colliding clusters. The gas provides evidence that a collision took place.

significant, if not more so. That's because this thin, tenuous gas, which is millions of degrees hot and pervades the space around the galaxies, actually has more mass than all of the galaxies put together.

When you add the galaxies and the gas up, you have about seventeen cents out of a galaxy cluster's dollar worth of mass. In other words, there is much more matter that we cannot see with telescopes but that we know is lurking within these clusters. This is the mysterious stuff called dark matter.

A GALAXY CLUSTER MASH-UP

By now, you may have noticed a pattern: things in the Universe on every scale run into each other. Galaxy clusters, despite their enormous size, are no different. When two galaxy clusters do collide, it is quite an event. While the individual galaxies in the cluster may pass through each other unscathed—remember, galaxy clusters are incredibly big, so there is a lot of space in between the galaxies—the hot gas does not.

Galaxy cluster Abell 520 (composite of X-ray, optical, and mass).

Instead, these two clumps of superheated gas slam into each other and become even more energized. The dark matter within the clusters comes along for the mash-up, and this gives astronomers an opportunity to look at what happens when it does.

DARK ENERGY+
DARK MATTER=
DARK MYSTERIES

How are things as mammoth as galaxy clusters created? Astronomers think it happens over a slow buildup of about a billion years. During this time, clumps of dark matter are drawn together through their massive gravitational strength, pulling galaxies with them. At first, small groups of galaxies form. Over much more time, these groups coalesce, and ultimately the bigger and biggest galaxy clusters form.

This illustration shows individual galaxies in white and yellow, with the gas between the galaxies shown in red. Galaxy clusters contain as much mass as a million billion suns. The gas is super hot—about a million degrees—so it can be seen as X-ray light.

It may be tough enough to contemplate how undetected, unseen material like dark matter affects some of the largest things in the Universe. But hang on—because here's where things get even trickier. The formation of galaxy clusters is not dependent on just dark matter. It is also influenced by the presence of something called dark energy.

The topic of dark energy deserves a little history. Back in 1998, a couple of groups of astronomers were studying very distant supernovas. More specifically, they were looking at one type of supernova that is thought to explode in the exact same way with the same brightness every time. (See Chapter 4 for more information on supernova types.)

If you know how inherently bright something is, then you can tell how far away it is based on how bright it looks. Think of a standard 60-watt light bulb. If you hold that at arm's length, then it's bright. If you place it across the street, it looks dimmer. And if you saw that light bulb as a tiny pinpoint of light, you would determine that the light bulb was pretty far away.

That's basically what the astronomers were doing with these supernova explosions, except they were using a lot more math. Because these supernovas are so bright, they can be seen at really far distances—billions of light-years away. The hope was to track how the Universe was evolving over its lifetime.

Almost every astronomer believes that the Universe started with the Big Bang about 13.7 billion years ago. From this monumental event, everything started flinging outward into what became the Universe. This "flinging outward" part is the key. Astronomers expected the Universe to slow its outward expansion after several billion years and maybe even come to a complete stop or start to contract a little. It's kind of like a ball being tossed up into the air. No matter how hard you throw the ball, sooner or later it will to start to slow its movement upward before eventually falling back toward the Earth because of gravity.

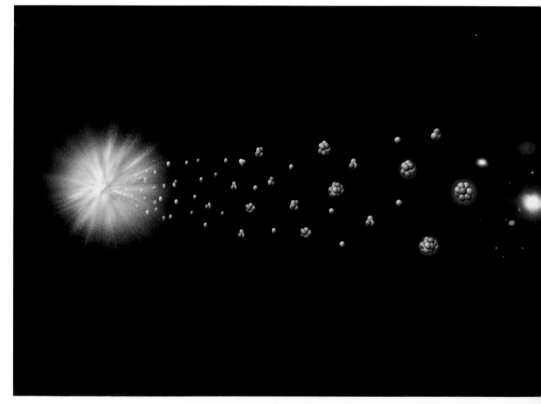

A simplified illustration of the history of our Universe over the past 13.7 billion years: starting on the left is the Big Bang. After the Big Bang, the first tiny bits of matter formed, eventually leading to the first stars (around 300-400 million years later), the first galaxies (about 1 billion years after the Big Bang), and eventually to large clusters of galaxies.

Quite often, the question of whether something is worthy of a Nobel Prize takes decades to be resolved. After all, many of the recipients of the prestigious Nobel are in the twilight of their careers when they are awarded the prize for things they did when they were much younger.

So it might be telling just how big of a deal it was that the 2011 Nobel Prize in physics was awarded to the two teams who were responsible for dark energy's discovery just thirteen years after it was made. We suppose that's what happens when you shake the very foundation of how the Universe works. We're also willing to go out on a limb and predict a Nobel for the first person or team who can actually tell us what dark energy is.

This is what the supernova-seeking astronomers were trying to measure. Instead of finding out that the Universe at this mature stage was slowing down, however, they found something completely shocking. They discovered that the Universe's expansion was actually accelerating. Imagine an explosion that doesn't end, but instead picks up speed and keeps going. That's like throwing our ball up, but instead of coming back down to Earth, it picks up speed and keeps flying away into the sky. These new results implied this was happening with our entire Universe.

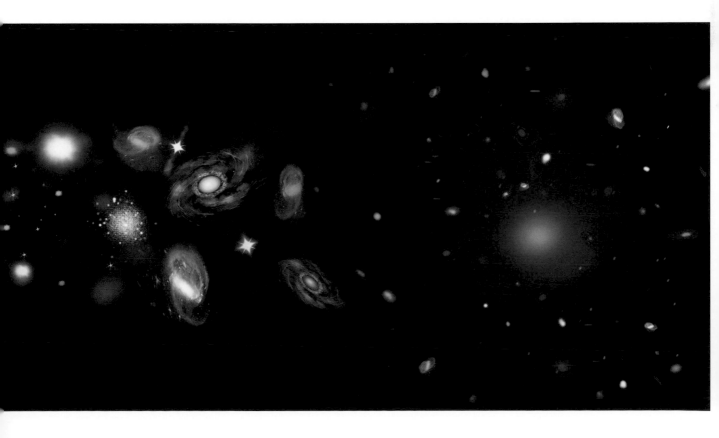

For a while, this was too strange for even astrophysicists to digest. They looked for problems in the studies. They tried to figure out if there was something that wasn't well understood about the supernovas themselves. But the more they looked, the more it appeared that the phenomenon was real. Eventually, scientists gave this mysterious force that was pushing the Universe apart at an accelerating rate the name of "dark energy." While they both share the word "dark," dark matter and dark energy are very different. Scientists think dark matter is stuff, while dark energy is a form of energy we don't yet understand. Some have called it an "antigravity" for its apparent effects on the Universe, but not every astronomer agrees it's the best description.

What does this have to do with galaxy clusters? Astronomers now think that the rate of how quickly galaxy clusters grow has to do not only with how much dark matter there is, but also with the presence of dark energy. If dark energy is pushing things apart, then giant clusters should grow more slowly as the expansion of the Universe accelerates. Astronomers have tested this by studying galaxy clusters in the distant Universe, finding that, yes, it looks like dark energy is really there. Galaxy clusters are, in some ways, the very big canary in the coalmine. Because they are so massive, the subtle effects of dark matter and dark energy have a relatively big imprint on them. So astronomers are interested in galaxy clusters not just because they are gigantic, but also because they hold many of the clues to some of the biggest questions in the Universe.

A CLUSTER OF GALACTIC GOODIES

GALAXIES ARE SOCIAL AND ARE OFTEN FOUND IN LARGE GROUPINGS CALLED CLUSTERS.

GALAXY CLUSTERS ARE SOME OF THE LARGEST OBJECTS IN THE ENTIRE UNIVERSE.

SCIENTISTS USE GALAXY CLUSTERS TO STUDY TWO OF THE BIGGEST MYSTERIES IN ALL OF SCIENCE: WHAT IS DARK MATTER, AND WHAT IS DARK ENERGY?

THE ANSWERS TO THESE QUESTIONS WILL DETERMINE THE FATE OF THE UNIVERSE AS WE KNOW IT.

STEPHAN'S QUINTET: 280 MILLION LIGHT-YEARS

Stephan's Quintet is a relatively small group made up of a handful of galaxies. This composite image of X-rays from Chandra (light blue) and optical data from the Canada-France-Hawaii Telescope (yellow, red, white, and blue) shows a beautiful look at it. Scientists think one galaxy is passing through the others at almost two million miles per hour. This generates a shock wave that heats the gas and creates the ridge of X-ray emission detected by Chandra.

ABELL 3376: 614 MILLION LIGHT-YEARS

This image of galaxy cluster Abell 3376 shows X-ray data in gold, as well as optical and radio data (the galaxies visible in the field). Scientists use observations of galaxy clusters such as Abell 3376 to study the properties of gravity on cosmic scales and test Einstein's theory of General Relativity. Studies like these are crucial for understanding the evolution of the Universe, both in the past and the future, and for studying the nature of dark energy, one of the biggest mysteries in science.

EL GORDO: 7.2 BILLION LIGHT-YEARS

A good nickname is worth a lot in astronomy. This galaxy cluster has been dubbed "El Gordo" for the "big" or "fat" one in Spanish. The name fits this galaxy cluster, which appears to be the most massive and the hottest and to give off the most X-rays (blue) of any known galaxy cluster at its distance or beyond. The comet-like shape of the X-rays, along with optical data, show that El Gordo is actually the site of a collision between two galaxy clusters. El Gordo was discovered using a telescope in Chile, which helped influence its excellent nickname.

ABOVE LEFT:

VIRGO CLUSTER OF GALAXIES: 48 MILLION LIGHT-YEARS

There are at least a thousand member galaxies that make up the Virgo Cluster of galaxies, which covers a large area on the sky. This mosaic of images shows the central region of the Virgo Cluster through faint foreground dust that lies above the plane of our own Milky Way galaxy. Since galaxy clusters are immense, some individual galaxies are a little farther or nearer to us. But on average, the galaxies in the Virgo Cluster are thought to be about 48 million light-years away.

CL 0024+17: 5 BILLION LIGHT-YEARS

The ring of dark matter (blue) in the galaxy cluster Cl 0024+17 is a strong piece of evidence for the existence of dark matter, an unknown substance that pervades the Universe. The ring was found by examining Hubble observations to see how the gravity of the cluster distorts the light of more distant galaxies (an effect called gravitational lensing.) In this way, astronomers can infer the existence of dark matter even though they cannot directly see it. Astronomers think that the dark matter ring was produced from a collision between two gigantic clusters.

MACSJ0717.5+3745: 5.4 BILLION LIGHT-YEARS

MACSJ0717.5+3745 is another messy galaxy cluster. MACSJ0717, for short, has four separate galaxy clusters involved in a collision. In this composite image, data from the Chandra X-ray Observatory reveal the cluster's hot gas, while an optical image from the Hubble Space Telescope shows the individual galaxies in the system. The hot gas in this image is color-coded to show temperature, where the coolest gas is reddish purple, the hottest gas is blue, and the temperatures in between are purple.

1E 0657-56: 3.8 BILLION LIGHT-YEARS

This galaxy cluster has an official name of 1E 0657-56, but it is much better known as the "Bullet Cluster." It earned its name because it contains a spectacular bullet-shaped cloud of hundred-million-degree gas that formed after the collision of two large clusters of galaxies. Hot gas detected by Chandra is seen as two pink clumps in the image and contains most of the "normal" matter in the two clusters. An optical image from Magellan and the Hubble Space Telescope shows galaxies in orange and white. The blue clumps show where most of the mass in the clusters is found, using a technique known as gravitational lensing. Most of the matter in the clusters (blue) is clearly separate from the normal matter (pink). This gives direct evidence that nearly all of the matter in the clusters is, in fact, dark matter.

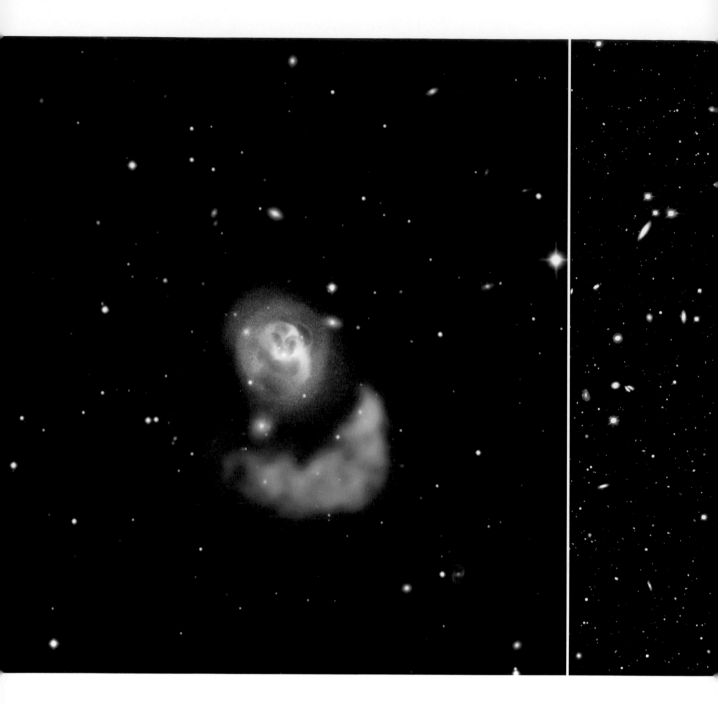

ABOVE LEFT:

ABELL 2052: 480 MILLION LIGHT-YEARS

Like wine in a glass, the hot gas in the galaxy cluster Abell 2052 is being sloshed back and forth. This huge spiral structure—seen in X-rays (blue)—was created when a small cluster of galaxies smashed into a larger one that surrounds a central elliptical galaxy. This sloshing has important effects including impacting how the giant elliptical galaxy and its supermassive black hole grow.

GRAVITATIONAL LENSING: 5 BILLION LIGHT-YEARS

Albert Einstein predicted many things—and he was right about many of them, too! One of them was that a mass that was big enough could act as a type of cosmic lens by distorting light from more distant objects. This Hubble Space Telescope image shows how a dense cluster of galaxies is bending the light from even more distant galaxies, leaving them looking like blue streaked arcs in this image. Studying the shapes and positions of these images, astronomers find there isn't enough visible matter to account for the distortions, so there must be a large amount of invisible dark matter present.

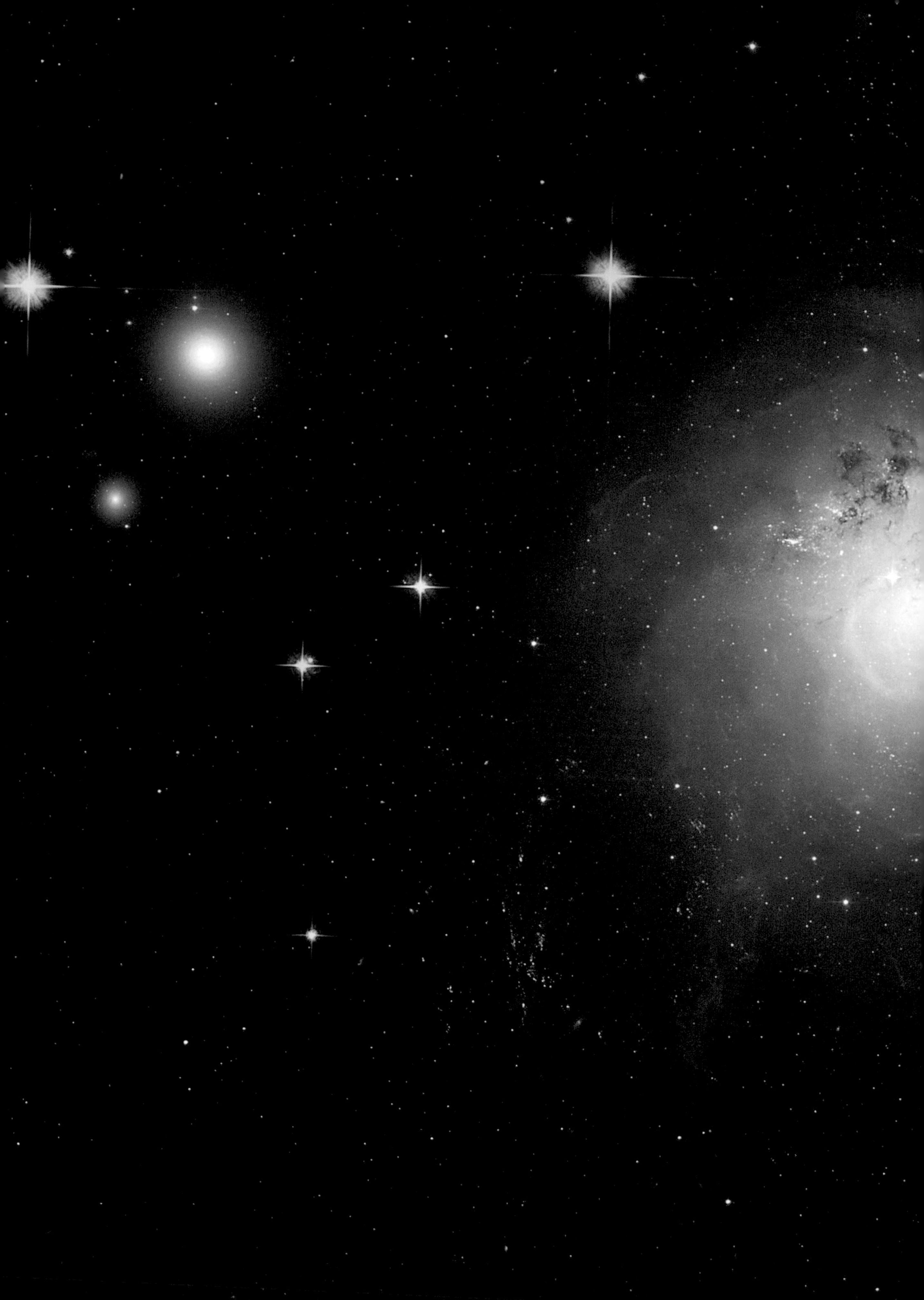

PERSEUS A: 250 MILLION LIGHT-YEARS

A galaxy called NGC 1275 lies at the center of the cluster of galaxies known as the Perseus Cluster. By combining different types of light into one image, scientists can see the dynamics of the galaxy more easily. In this composite image, X-rays from Chandra are shown in violet and reveal the presence of a black hole at the center of NGC 1275. Optical data from Hubble is depicted in red, green, and blue, and radio emission in pink traces the jets generated from the central black hole.

9

TAKE THE LONG
WAY HOME

For my part I know nothing with any certainty but the sight of the stars makes me dream.

— VINCENT VAN GOGH

AS WE NEAR THE END OF OUR JOURNEY, LET'S BEGIN with three observations:

This book doesn't include a complete, tidily wrapped up story.

There are many, many mysteries and questions to explore throughout and about the Universe.

The quest to investigate these questions is ongoing, and there is no telling what the answers will look like.

We have traveled from the very familiar locale of our home planet to structures that are so colossal that they almost defy our understanding. During this trip, we have gone through enormous leaps in distance and scale, while experiencing phenomena that almost seem too strange and fascinating to be true.

The amazing thing about studying the Universe is this: while some theories and ideas may change as scientists gather new data, the whole thing just gets more interesting.

PREVIOUS SPREAD:

This image is the farthest we have yet been able to see in visible light. The Hubble Space Telescope recorded objects that are a hundred million times fainter than can be seen with the naked eye. The image shows some of the most distant galaxies as they were only a few hundred million years after the Big Bang.

THE SEARCH FOR LIFE

For many people, the fun of exploring the cosmos is not only finding answers but also revealing new questions. One of the most intriguing questions out there is one that many of us have asked: Are we alone? It seems like there are practically daily reports of planets and planetary systems being found that inch closer and closer to resembling Earth and its Solar System. Is there one out there that truly mirrors our own home planet? Are there other planets that are truly hospitable to life like ours? Of course, the big elephant in the room is, if there are other types of life in this vast Universe, where are they?

OPPOSITE:

This artist's drawing illustrates one of the many planet-star systems discovered by NASA's Kepler mission. We can't directly image them yet, but scientists study their combined light and how it behaves to characteristics.

UNIVERSAL QUESTIONS

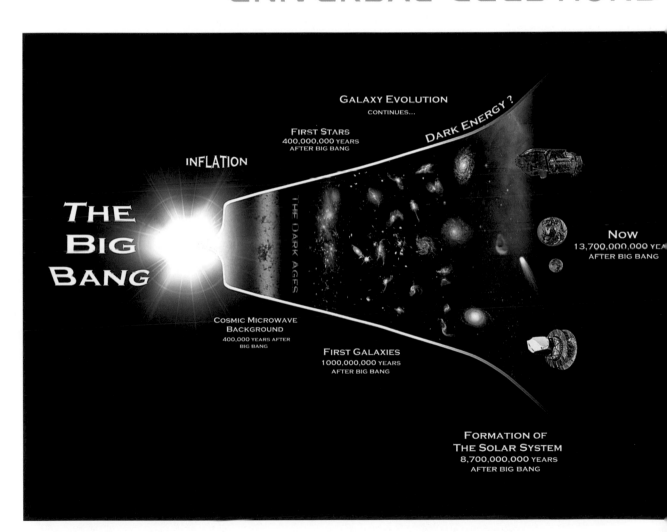

GALAXY EVOLUTION
CONTINUES...

FIRST STARS
400,000,000 YEARS
AFTER BIG BANG

DARK ENERGY?

INFLATION

THE BIG BANG

THE DARK AGES

NOW
13,700,000,000 YEARS
AFTER BIG BANG

COSMIC MICROWAVE
BACKGROUND
400,000 YEARS AFTER
BIG BANG

FIRST GALAXIES
1,000,000,000 YEARS
AFTER BIG BANG

FORMATION OF
THE SOLAR SYSTEM
8,700,000,000 YEARS
AFTER BIG BANG

This illustration shows the evolution of our Universe starting with the Big Bang. Moving from left to right shows the advancement of time since the earliest days of the Universe. Scientists think the first stars formed a few hundred million years after the Big Bang, the first galaxies formed not much after that, and our Solar System came into being about 9 billion years after the Big Bang.

While most of wait and some of us look for an answer to that question, there are so many more mysteries to ponder. We talked about how the Universe is thought to have started in the Big Bang nearly 14 billion years ago. From there, it rapidly expanded as a superheated sea of hydrogen and helium atoms. During that time, there is a period astronomers refer to as the "Dark Ages" because we cannot gather any information on it. What happened then? Once the Universe cooled enough after about a few hundred million years the first stars and galaxies were able to form. How exactly did this happen?

The consensus is that these seedling galaxies merged and grew, eventually growing into the mature and complex galaxies we see (and live in) today. Did the black holes in the centers of these galaxies form first or vice versa? If they evolved together, what exactly does a codependent relationship between a black hole and a galaxy look like?

Across the Universe, we see major mysteries including those of dark matter and dark energy. In fact, adding up what we can directly detect with all of our highly sophisticated telescopes, we can only account for a paltry 4 percent of the Universe. To put it another way, as much as we do know, there is about 96 percent of the Universe that we have no clear way to explain.

BELOW:

We've never been big fans of licorice-flavored jelly beans, but putting that aside, this graphic helps show just how little normal matter there is in the Universe. The Universe is mostly dark. Only a little over 4 percent of the Universe is the stuff that makes up stars, planets, and people. The rest of the Universe, about 96 percent, consists of dark energy and dark matter.

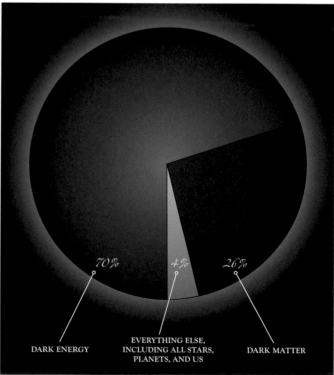

70%

4%

26%

DARK ENERGY

EVERYTHING ELSE, INCLUDING ALL STARS, PLANETS, AND US

DARK MATTER

There are many ways to access the plethora of gorgeous space images that are available today. You can pick up a book (like this one, of course), or look at them on your phone, tablet, computer, or other favorite electronic device.

However, we believe that there is something to be said for seeing these images in different environments, in person. Not only are the images displayed on larger scales in exhibits, festivals, or museums, but they are also sometimes placed in contexts that make us think about what we are seeing a little bit differently.

In the past several years, there has been a surge in science festivals across the United States. These events, which have been popular in Europe and elsewhere around the world for decades, can now be found annually in cities like Washington, D.C., New York, and San Diego. Often, space agencies like NASA will have booths at these free public events where you can look at the latest and greatest from space and have the chance to ask a question or two.

There are also projects that specifically put astronomy into the public sphere. Two of those projects include the "From Earth to the Universe" and "From Earth to the Solar System" (materials from both of these projects were adapted for this book). You can look for a calendar of where these ongoing projects are on display through their websites at www.fromearth-totheuniverse.org and fettss.arc.nasa.gov.

SEEING SPACE UP CLOSE

STUDYING FROM HERE TO LEARN ABOUT THERE

How will we get to the bottom of these giant mysteries? Astronomers are looking for clues across the Universe. As we speak, there are teams of scientists and engineers located in facilities around the world that are planning new telescopes and observatories to probe these questions in different ways. One thing is certain: whenever we have successfully built and operated a new telescope, wonderful and unanticipated discoveries are made.

At the same time, particle physicists have constructed giant accelerators here on Earth, such as the recently completed Large Hadron Collider in Switzerland. Monumental facilities like this and other complex detectors built on our planet may just reveal previously unknown particles that could lead to new insights about some of these puzzles—like the nature of dark matter—from the cosmos.

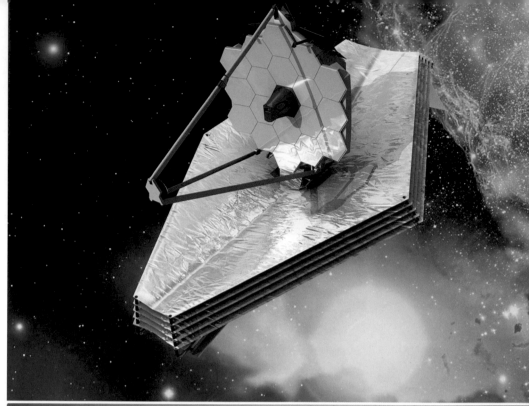

This artist's illustration shows the upcoming James Webb Space Telescope (JWST), a planned space telescope optimized for observations in the infrared, as a successor to NASA's Hubble Space Telescope and Spitzer Space Telescope.

The accelerator ring of the Large Hadron Collider as photographed from the air.

OPPOSITE:
The almost-full Moon rises, and the night sky beckons.

It's fitting that some of the answers we're looking for might be found—at least in part—right here on Earth. It's a reminder that we are all part of one interconnected Universe. Questions from "out there" may be answered from here, and the reverse is likely also true. Who knows what kind of knowledge we may find about our very existence on this planet by stretching our minds and exploration across the vast distances of the Universe?

WELCOME BACK HOME

All travel experiences must end, and there's always a point when it's time to head for home. But this journey is unlike most that we take, because in the end, we've actually never left. We have found, however, that the fascinating phenomena of our Solar System, our Galaxy, our Universe as a whole, are all part of the same web of wonder into which all of us are interwoven.

This book, as all travel guides are, is incomplete. It provides a mere sampling of places and things to explore in our Universe. If your appetite for the cosmos has been whetted, we hope you'll continue your investigation. Luckily, it requires little more than a trip to your local library, bookstore, or nearest computer. So get on out there—the Universe is yours to discover.

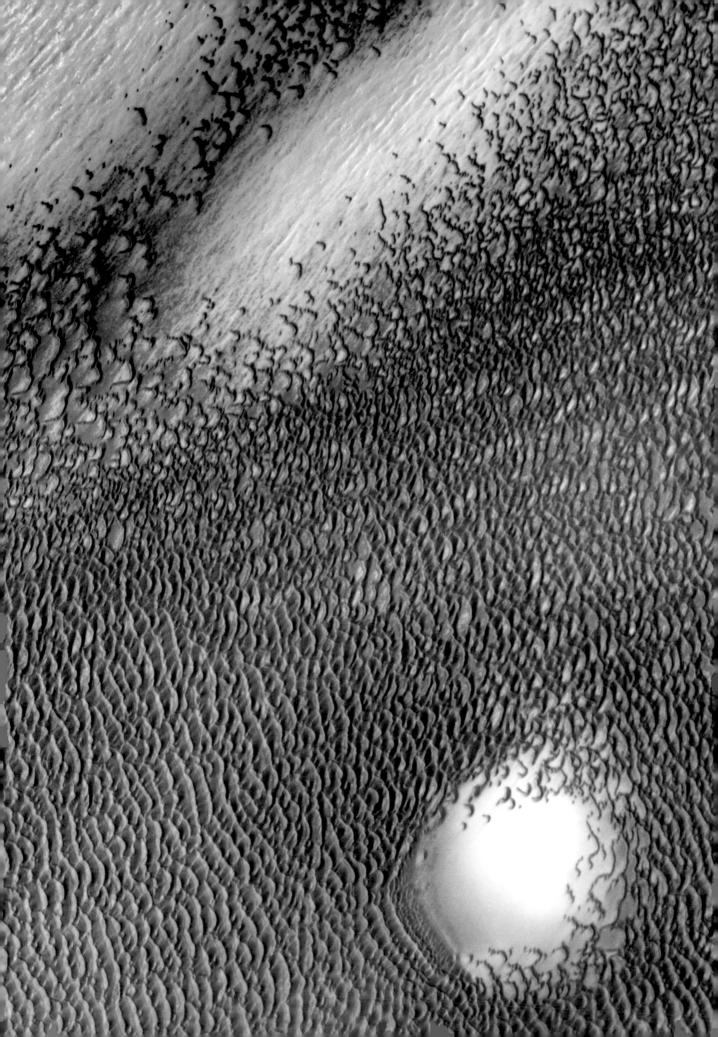

BOOKS

Jim Bell, *Postcards from Mars: The First Photographer on the Red Planet* (Plume, 2010)

Stephen Hawking, *The Universe in a Nutshell* (Bantam, 2001)

Alan Lightman, *Einstein's Dreams* (Knopf, 1994)

Mario Livio, *The Accelerating Universe* (John Wiley & Sons, 2000)

Philip Plait, *Bad Astronomy: Misconceptions and Misuses Revealed, from Astrology to the Moon Landing "Hoax"* (Wiley, 2002)

Mary Roach, *Packing for Mars: The Curious Science of Life in the Void* (Norton, 2011)

Carl Sagan, *Cosmos* (Harper & Row, 1974)

Neil deGrasse Tyson, *Death by Black Hole: And Other Cosmic Quandaries* (Norton, 2007)

WEBSITES

Astronomy Picture of the Day (apod.nasa.gov)

Bad Astronomy Blog (blogs.discovermagazine.com/badastronomy)

Chandra X-ray Observatory (chandra.si.edu)

Earth Observatory (earthobservatory.nasa.gov)

European Space Agency (esa.int)

From Earth to the Solar System (fettss.arc.nasa.gov)

From Earth to the Universe (fromearthtotheuniverse.org)

Hubble Space Telescope (hubblesite.org)

NASA (nasa.gov)

Portal to the Universe (portaltotheuniverse.org)

Solar System Exploration (solarsystem.nasa.gov)

Space.com (space.com)

Visible Earth (visibleearth.nasa.gov)

World Wide Telescope (worldwidetelescope.org)

OPPOSITE:
Did you know that sand dunes can be found across Mars? Just as on Earth, strong winds blow and help shape the planet's surface. The sea of dunes in this image was sculpted by Martian wind into long lines. The dunes surround the northern polar cap of Mars, covering an area as big as Texas. In the color-coded image, blue shows the cooler regions, while the warmer regions are yellow and orange.

ACKNOWLEDGMENTS

THERE ARE MANY PEOPLE THAT WE WOULD LIKE to thank for the completion of this book. We would like to acknowledge our agent, Elizabeth Evans (Jean V. Naggar Literary Agency), who approached us with the idea to do a book in the first place. We thank Carolyn Gleason, director at Smithsonian Books, and her staff for helping guide us through our first book. We also thank our editor, Gregory McNamee.

The genesis of this book was our project for the International Year of Astronomy (IYA) in 2009 called "From Earth to the Universe" (FETTU). FETTU was an incredibly enriching project for both of us professionally and personally. We thank Lars Lindberg Christensen (European Southern Observatory, ESO) and Pedro Russo (Leiden University) for their tremendous support of FETTU and their vision for IYA. Gary Evans (Science Photo Library) was instrumental in getting FETTU launched. We also thank Manolis Zoulias (National Observatory of Athens), Jean-Charles Cuillandre (Canada-France-Hawaii Telescope Corporation), Thierry Botti (OAMP), Rick Fienberg (AAS), Henri Boffin (ESO), and Antonio Pedrosa (Navegar Foundation), who also worked on the FETTU task group that helped curate the original FETTU materials, many of which were adapted for this book.

"From Earth to the Solar System" (FETTSS) gave us a second wind. We thank Daniella Scalice and Julie Fletcher (NASA Astrobiology Institute) for launching that idea and for co-curating the FETTSS materials that were also adapted for this book.

For all the artists and astrophotographers who volunteered the use of their work in the FETTU and FETTSS projects, as well as this book, we are ever grateful. Your images helped make the project a success.

Special thanks go to Steve Lubar (Brown University) for his immense patience in reviewing this book. Peter Edmonds (Chandra X-ray Center, CXC) went above and beyond to help make sure that we are not only just mildly entertaining in our writing, but also accurate. Melissa Weiss (CXC) is unmatched in her eye for design, and we thank her for her support. Henri van Bentum was kind enough to catch last-minute typos in the text.

We would like to give thanks and credit to our co-workers at the Chandra X-ray Center, the X-ray light of our lives. We would not be writing this book at all if it were not for the expertise we developed while working on this fascinating mission (and we have picked out our favorite Chandra materials to share in this book). For that experience, we deeply thank Kathy Lestition (CXC) for all of her support, kindness, and consideration and Harvey Tananbaum (CXC) for his encouragement and kind words. The whole Chandra Education and Public Outreach (EPO) group has been a substantial part of our lives for some time, including Wallace Tucker, April Hobart, John Little, Kayren Phillips, Khajag Mgrdichian, Aldo Solares, and Lisa Portolese (CXC). Joe DePasquale (CXC), and Eli Bressert (CXC), you'll find that this book contains quite a few of the beautiful Chandra images you worked on.

Our professional counterparts at Hubble and Spitzer will find their work sprinkled throughout this book. We thank Zolt Levay and Lisa Frattare (Space Telescope Science Institute, STScI), Robert Hurt (JPL-Caltech) specifically, and the Hubble and Spitzer EPO offices.

We greatly appreciate Mario Livio (STScI) for taking the time to write the foreword. He is an inspiration for our work.

On a bigger scale, we would like to acknowledge both NASA and the Smithsonian Institution. We are grateful for all of the support and inspiration that everyone we work with has provided. That support has been instrumental in inspiring us to talk about the Universe.

Finally, we would like to dedicate this book to our families and friends who put up with us while writing it. All of you make living here on Earth completely worthwhile.

Kim: My husband, John, was so helpful with everything I neglected while writing this. I truly couldn't have done it without him. My children, Jackson and Clara, offered much patience and love while I was "busy with the book." My mother, Chris, my biggest fan, supported me every day. My mother-in-law, Diana, helped out even after the loss of my dear father-in-law, John. My father, John, brother Scott, step-mother Vicki, siblings-in-law, and all the rest of my family, have been cheerleaders through and through.

Megan: Thank you to Kristin, Anders, Jorja, and Iver, who motivate and inspire me to do so many things. I am grateful to my parents, who taught me to do what I love even if there is no obvious benefit in sight. To the rest of my family and friends, I hope this book helps finally explain a little bit more about "what I do."

B=Bottom; C=Center; L=Left; R=Right; T=Top

2MASS: 158-159T (J. Carpenter, M. Skrutskie, & R. Hurt)
Adam Block and Tim Puckett: 137
AMASE: 20-21 (Kjell Ove Storvik)
American Museum of Natural History: 52 (David Hardy)
Ariel Anbar: 24
© Australian Astronomical Observatory: 2-3 (David Malin), 139 (David Malin), 174 (David Malin)
AVHRR/NDVI/Seawifs/MODIS/NCEP/ DMSP/Sky2000 star catalog: 28-29 (AVHRR & Seawifs texture by Reto Stockli, visualization by Marit Jentoft-Nils)
California Association For Research In Astronomy/Science Photo Library: 87
Canada-France-Hawaii Telescope Corporation: 8 (J.C. Cuillandre), 193 (J.C. Cuillandre & G. Anselmi), 215 (Coelum), 218 (J.C. Cuillandre)
Cardiff U.: 229 (Rhys Taylor, www.rhysy.net)
CERN: 232B
Ciel et Espace: 63 (Akira Fujii), 64 (Jean-Luc Dauvergne)
Dan Schechter: 112
Darren Edwards and Suzi Taylor: 17
Dominion Radio Astrophysical Observatory: 171 (Jayanne English & Tom Landecker)
Eckhard Slawik: 73, 118, 166-167
ESA: 37BR, 232T (C. Carreau)
ESA/DLR/Freie Universität Berlin: 13T (G. Neukum)
ESA/VIRTIS/INAF-IASF/Observatory de Paris-Lesia: 79
ESO: 120T (L. Calçada), 125 (S. Steinhöfel), 183
ESO/IDA: 192 (R. Gendler & C. Thöne)
ESO/VISTA: 143 (J. Emerson)
ESO/VLT: 220
ESO/VLT/Pontificia Universidad Catolica de Chile: 217 (L. Infante)
Fermilab: 230BL
Gemini Altair Team/NOAO/AURA/NSF: 106 (Travis Rector & Chad Trujillo)
Gemini Observatory/U. of Alaska Anchorage: 194 (Travis Rector)
Greg Piepol: 72
Henry Bortman: 27
High Lakes Project/NASA Astrobiology Institute/SETI CSC/NASA Ames Research Center: 22, 23
HST WFC3 Science Oversight Committee/NASA/ESA: 121 (F. Paresce & R. O'Connell)
ISS Crew Earth Observations Experiment and NASA/JSC/Image Science and Analysis Laboratory: 47
Jack Newton: 91, 71
Jenn Macalady, Becky McCauley, and Hiroshi Hamasaki: 25
Johannes Schedler: 136
Jordi Gallego: 208
Kerry-Ann Lecky Hepburn: 150-151
Magellan/U. Arizona: 221 (D. Clowe et al.)
Mark Boyle: 18
MIT/NASA Astrobiology Institute: 26 (Phoebe Cohen)
MPIA, Calar Alto: 145 (O. Krause et al.)
NAOJ/Subaru: 220
NASA: 6, 44, 51, 56T, 85, 103
NASA Goddard Space Flight Center/ NOAA/NGDC: 40-41 (Christopher Elvidge, Marc Imhoff, Craig Mayhew, & Robert Simmon)
NASA GOES Project Science/NOAA Comprehensive Large Array-Data Stewardship System: 43
NASA Johnson Space Center: 37RT

NASA Kepler Mission: 55 (Jason Rowe)
NASA/ACS Science Team/ESA: 164 (H. Ford, G. Illingworth, M. Clampin, & G. Hartig)
NASA/Ames/JPL-Caltech: 228
NASA/CXC: 30, 32 (M.Weiss), 76 (M.Weiss), 120B (S. Lee), 122 (M.Weiss), 124 (K. Arcand), 129 (A.Hobart), 156 (M.Weiss), 158B (M.Weiss), 160 (M.Weiss), 162B (M.Weiss), 165 (M.Weiss), 206-207 (M.Weiss), 211 (A.Hobart), 212-213 (M.Weiss), 230BR (M.Weiss)
NASA/CXC/CfA: 185 (D. Evans et al.), 198 (R. Kraft et al.), 215 (E. O'Sullivan), 221 (M. Markevitch et al.)
NASA/CXC/GSFC: 133 (M. Corcoran et al.)
NASA/CXC/IoA: 224-225 (A. Fabian et al.)
NASA/CXC/ITA/INAF: 220 (J. Merten et al.)
NASA/CXC/JHU: 195 (D. Strickland)
NASA/CXC/KIPAC: 196 (S. Allen et al.), 197 (N. Werner & E. Million et al.)
NASA/CXC/MIT/UMass Amherst: 144 (M.D. Stage et al.)
NASA/CXC/MIT: 142 (D. Dewey et al.), 170 (F.K. Baganoff et al.), 203 (C. Canizares & M. Nowak)
NASA/CXC/NCSU: 127TL-TR (S. Reynolds et al.)
NASA/CXC/Penn State: 140 (L. Townsley et al.), 146 (S. Park et al.)
NASA/CXC/RIT: 123 (J. Kastner et al.)
NASA/CXC/ROSAT/Asaoka/DSS: 128 (B. Gaensler & Aschenbach)
NASA/CXC/Rutgers: 217 (J. Hughes et al.)
NASA/CXC/S.Wolk: 33 (modified by K. Arcand)
NASA/CXC/SAO: 66 (J. Drake & Robert Gendler), 127BL (F. Seward, modified by K. Arcand), 142 (J. DePasquale), 145, 147, 148-149, 199 (J. DePasquale), 216 (A. Vikhlinin)
NASA/CXC/UMass: 34 (D. Wang et al.), 168-169 (D. Wang et al.)
NASA/CXC/U. Waterloo: 222 (B. McNamara)
NASA/ESA: 89L (M. Buie), 111 (L. Lamy), 154 (E. Jullo, P. Natarajan, & J.P. Kneib), 177 (A. Nota), 187 (Peter Anders)
NASA/ESA/ASU: 127BR (J. Hester & A. Loll)
NASA/ESA/CFHT/CXO: 209TL-BL (M.J. Jee & A. Mahdavi), 210 (M.J. Jee & A. Mahdavi)
NASA/ESA/Hubble Heritage (STScI/ AURA) & U. Cambridge/IoA: 224-225 (A. Fabian)
NASA/ESA/Hubble Heritage Team: 135, 182, 184, 186, 200 (S. Beckwith), 202 (P. Goudfrooij), 226-227 (S. Beckwith)
NASA/ESA/Hubble Heritage Team (STScI/AURA): 147, 172, 173, 180-181, 195
NASA/ESA/JHU: 223 (M. Lee & H. Ford)
NASA/ESA/McMaster U.: 196 (W. Harris)
NASA/ESA/STScI: 168-169 (D. Wang et al.)
NASA/ESA/STScI/U. Waterloo: 222L (B. McNamara)
NASA/Goddard/Arizona State U.: 65
NASA/GSFC/EOS: 31
NASA/IAS: 189 (J. Bahcall)
NASA/JHU Applied Physics Laboratory/ Carnegie Institution of Washington: 61, 78, 96
NASA/JPL: 88
NASA/JPL/DLR: 108TL,TR,BL,BR
NASA/JPL/Penn State: 140 (L. Townsley et al.)
NASA/JPL/Rutgers: 217 (F. Menanteau)
NASA/JPL/Space Science Inst.: 109,

110
NASA/JPL/Texas A&M/Cornell: 74-75
NASA/JPL/U. of Arizona: 84, 93, 94, 95, 100, 104
NASA/JPL/Voyager 2 Team: 107
NASA/JPL-Caltech: 82, 89R (R. Hurt, modified by K. Arcand), 97, 105T-B, 145, 199
NASA/JPL-Caltech & SAGE Legacy Team: 175 (M. Meixner)
NASA/JPL-Caltech/Arizona State U./ THEMIS: 80-81, 234
NASA/JPL-Caltech/GALEX Team: 194 (J. Huchra et al.)
NASA/JPL-Caltech/SSC: 168-169 (S. Stolovy)
NASA/JPL-Caltech/STScI: 176
NASA/JPL-Caltech/U. Arizona: 101, 102, 195 (C. Engelbracht)
NASA/JSC/Earth Sciences and Image Analysis Laboratory: 16, 39, 42 (Mike Trenchard), 45, 46, 48-49
NASA/Mars Global Surveyor: 98
NASA/NOAA/GSFC/Suomi/NPP/VIIRS: 38 (Norman Kuring)
NASA/SDO: 68, 69, 117
NASA/SOHO/EIT: 59
NASA/Stanford-Lockheed Institute for Space Research's TRACE Team: 58B, 70
NASA/STScI: 142, 148-149, 185, 199, 203, 220 (R. Dupke), 221, 228
NASA/STScI/NRAO: 229 (A. Evans et al.)
NASA/STScI/U. Washington: 123 (B. Balick)
National Science Foundation: 15 (Joe Mastroianni)
NOAA: 36
NOAO/AURA/NSF: 113 (T. Rector, Z. Levay & L. Frattare), 132 (Jay Ballauer & Adam Block), 138 (T. Rector & B. Wolpa), 141 (Nathan Smith)
Nobeyama Observatory: 57L
NRAO/AUI: 86
NRAO/AUI/NSF: 197 (F. Owen)
NRAO/Ohio U.: 57L (L. Birzan et al.)
NRAO/VLA: 196 (G. Taylor), 224-225 (G. Taylor)
NSF/NRAO/VLA/IUCAA: 216 (J. Bagchi)
NSF/VLA/CFA: 185 (D. Evans et al.)
Palomar Observatory Sky Survey: 134 (Davide De Martin), 146
R. Jay GaBany: 195
Robert Gendler: 178-179
Rogelio Bernal Andreo: 219
Russell Croman: 204-205
Science Photo Library: 58B (Pekka Parviainen), 67 (Laurent Laveder)
SDSS/Galaxy Zoo: 190 (Richard Nowell & Hannah Hutchins)
Shutterstock: 10-11 (Songchai W.)
SOAR/MSU/NOAO/UNC/CNPq Brazil/Rutgers: 217 (F. Menanteau)
SOHO/NASA/ESA: 56B, 57CR
SOLIS/NSO/AURA/NSF: 57CL
SSC/JPL/Caltech: 159B (Robert Hurt, modified by K.Arcand)
www.stargazer-observatory.com, http://www.astrogrossi.de/: 188 (© Dietmar Hager, F.R.A.S. & Torsten Brandes)
Stephane Guisard: 153
Terrastro.Com/Science Photo Library: 114-115 (Alex Cherney)
TOMS Science Team/NASA Goddard Space Flight Center: 37L
USGS/National Map Seamless Server: 53
Wayne England: 233
Wes Skiles/National Geographic Stock: 19
Wikimedia: 13B, 83 (Dave Jarvis, modified by K. Arcand), 116
Yohkoh Legacy Archive (http://solar. physics.montana.edu/ylegacy), Montana State University: 5

Text © 2013 by Kimberly K. Arcand and Megan Watzke

This book may be purchased for educational, business, or sales promotional use. For information, please write:

Special Markets Department
Smithsonian Books
P. O. Box 37012, MRC 513
Washington, DC 20013

Published by Smithsonian Books
Director: Carolyn Gleason
Production Editor: Christina Wiginton
Editorial Assistants: Jane Gardner and Ashley Montague

Edited by Greg McNamee
Designed by Bill Anton | Service Station

Library of Congress Cataloging-in- Publication Data
Arcand, Kimberly K.
Your ticket to the universe: a guide to exploring the cosmos / Megan Watzke and Kimberly Arcand.
   pages cm
ISBN 978-1-58834-375-8 (pbk.)
1. Astronomy. 2. Universe.
3. Universe. 4. Cosmology. I. Watzke, Megan.
II. Title.
QB43.3.A73 2013
523.1—dc23   2012029115

Manufactured in China
17 16 15 14 13   5 4 3 2 1